Coquih

A Guide to the North Cascade Mountains and the Nicola Valley

Second Edition

by
Murphy Shewchuk

Sonotek® Publishing Ltd.
P.O. Box 1752
Merritt, B.C. Canada V1K 1B8

www.sonotek.com

First printing: December, 1990.
Revised and reprinted: July, 1991.
Complete update: November, 1997.

Canadian Cataloguing in Publication Data

Shewchuk, Murphy.
Coquihalla country

Includes bibliographical references and index.
ISBN 0-929069-10-2

1. Outdoor recreation—British Columbia—Guidebooks. 2. British Columbia—History—Guidebooks. 3. British Columbia—Guidebooks. 4. Natural history—British Columbia—Guidebooks. I. Title.

GV191.46.B75S5 1997 917.11'72 C97-910938-8

SONOTEK® PUBLISHING LTD.
P.O. Box 1752,
Merritt, B.C.
Canada V1K 1B8

Telephone and facsimile: (250) 378-5930
E-mail: sonotek@sonotek.com
URL http://www.sonotek.com

Edited by Gill Foss. Photographs and maps by Murphy Shewchuk unless otherwise credited.
Printed in Canada.

Dedication

To Allyson Lynn and Michael Lucas.

Acknowledgements

Coquihalla Country started off as an update to *Exploring the Nicola Valley*, originally published in 1981 by Douglas & McIntyre. However, as I began to retrace the routes that I had travelled a decade ago and several times since, I discovered a great many changes and a wealth of new information about the past and present. *Exploring the Nicola Valley* was to become merely a foundation for a book that would go beyond anything I had ever done before. What began as a simple summer project quickly grew into a major effort requiring thousands of kilometres of travel, by vehicle and on foot, and hundreds of hours at the computer keyboard and in the photography darkroom.

Thanks to the encouragement of *BC Outdoors* editors Roegan Lloydd and Karl Bruhn, and former editors Mary Aikins, Henry Frew and George Will, some of the building blocks were in place in the form of previously-published *Backroads* articles.

Much new material surfaced, however, delaying the project as each answer prompted new questions. Nancy Wise, Sonotek's distributor, encouraged me to continue even though deadlines had come and gone. Cecil Simpson and Stan Parkinson of BC Parks plus Graham Seefeldt, Lorne Robertson and Dave Crack helped considerably with information and suggestions. Fellow hikers and back-country explorers, including Jean Nelson, Jack and Wilma Grant, Harriet Newhouse, Glen and Margaret Carlson, Elizabeth Salomon-de-Friedberg, Dave Brown, Andrew and Andrea Howes, my wife Katharine and my grand-daughter Bridget, and her friends Nola and Matthew, made the research trips much more worthwhile and interesting.

Thanks also must go to Rick Howie, a well-known Kamloops naturalist, for the special section on birdwatching along Highway 5A, between Merritt and Kamloops — certainly one of the more productive birding areas in the BC Interior.

I would especially like to thank Katharine for her proof-reading and editorial suggestions.

Murphy Shewchuk

•••

Coquihalla Country

A Guide to the North Cascade Mountains and the Nicola Valley

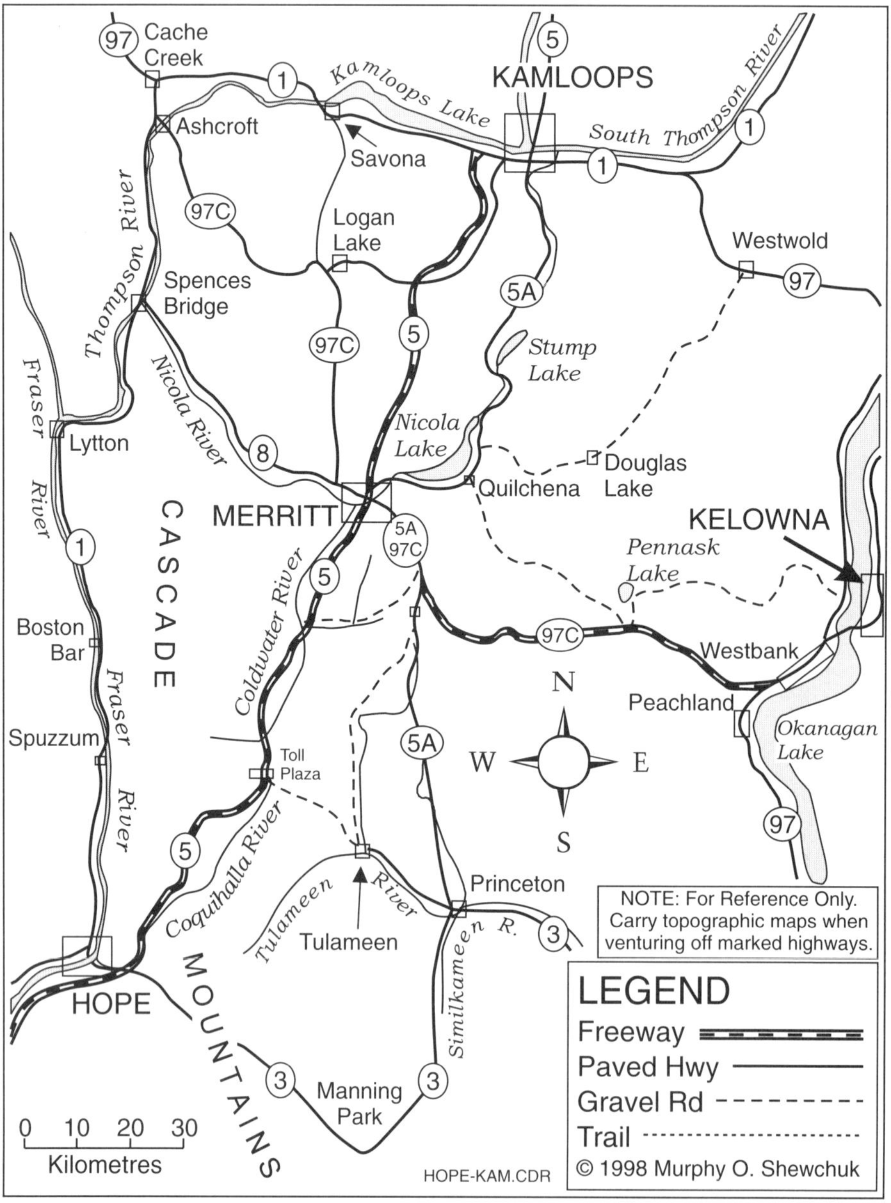

Map 1
Coquihalla Country

CONTENTS

Introduction

The heart of the Thompson Plateau and North Cascade Mountain region that has become known as *Coquihalla Country*, lies about three hour's drive northeast of Vancouver. It is bounded on the west and north by the Trans-Canada Highway and the Fraser and Thompson rivers; on the south by Crowsnest Highway 3 and the Similkameen River; and on the east by Highway 97 and the Okanagan Valley.

Coquihalla Highway 5 bisects the region from south to north. A second major highway, the Okanagan Connector (Highway 97C), ties the mid-point of the Coquihalla Highway to the Okanagan Valley. Highway 5A passes through the valley from Kamloops to Princeton, following the shorelines of many of the beautiful fishing lakes that lie along the lowland route. Highway 8 follows the Nicola River downstream from Merritt, through shade-cloth covered ginseng fields, hayfields, market gardens and a narrow, semi-desert valley before reaching Spences Bridge and the Trans-Canada Highway.

Highway 97C continues north from Highway 8 west of Merritt near the Native community of Shulus. Formerly the Mamit Lake Road, this route passes near the community of Logan Lake before winding northwest through the copper-rich Highland Valley and descending to Ashcroft on the Thompson River. To the north of the Logan Lake junction, a gravel road meanders through the lodgepole pine, past Tunkwa and Leighton lakes, before twisting down a steep hill into the village of Savona. To the east of Logan Lake, a paved route through the aspen and pine passes near Lac Le Jeune and descends to Kamloops on the Trans-Canada Highway. Countless dirt roads and trails branch off these more important routes.

A Canadian Pacific Railway (CPR) spur line, formerly part of the famous Kettle Valley Railway (KVR), once provided a freight route from Penticton to the main line at Spences Bridge via Princeton and Merritt. The service was discontinued in May, 1989. The rails have been removed, and gates and barriers established to limit vehicle traffic.

Hope, at the southern gateway to *Coquihalla Country*, is a mere 154 kilometres from Vancouver — 1.5 hours from western Canada's largest population centre. Merritt, a city of 7,000 and the largest centre within the region, is only 270 kilometres from Vancouver and 98 kilometres from downtown Kamloops.

The Thompson Plateau, which comprises the northern and eastern portions of *Coquihalla Country*, is a lake-dotted highland. The rounded, pine-covered mountains, open grasslands and many elongated lakes are clearly the result of ancient glacial action.

In contrast, the north slope of the Cascade Mountain chain, that makes up the southwestern quarter of the region, presents a startling change in topography and vegetation. Spire-like peaks reach for the sky. Cedar, spruce and hemlock fill the mountain valleys.

Anglers and hunters have long been aware of *Coquihalla Country's* many recreational attractions, but only in the past few decades have hikers, backpackers and cross-country skiers begun to explore the numerous trails. There is still much to be done before the vast potential of this rich and varied land is realized.

As a note of caution, the dry grasslands of the Nicola Valley are fragile and should be treated with respect. Please drive only on recognized roadways; the spring runoff or a sudden rainstorm can turn a hillside tire rut into a major gully, destroying the delicate rangeland.

Do not leave refuse behind. Stow it where containers are provided or take it out with you. A discarded bottle could start a fire in the hot summer sun or a discarded can could create a trap for wildlife.

•••

Important Notice!

Although every effort has been made to provide accurate information, road and trail conditions in *Coquihalla Country* are constantly changing. Consequently, neither the author nor the publisher can guarantee the continuing accuracy of this information.

We will, however, attempt to post any changes or corrections to our World Wide Web site at <http://www.sonotek.com> between printings.

We look forward to corrections and comments on ways that you think this book may be improved. Please write to the author care of the publisher at the address listed in the front of the book. You can also e-mail us at <publisher@sonotek.com>.

•••

Region in Transition

There have been major changes taking place with regard to recreation in the North Cascade Mountains and the Nicola Valley. When we set out to publish the Second Edition of *Coquihalla Country*, it was our plan to include new information on the former Kettle Valley Railway right-of-way and the proposed Trans Canada Trail route through the region. However, for numerous reasons, both subjects have remained impossible to pin down with any degree of accuracy.

It is known that the British Columbia government purchased most of the KVR right-of-way between Spences Bridge and Princeton (except for Indian Reserves). It appears to be their intention that the route will remain in public use. However, it has been very difficult to confirm much beyond that.

There has also been considerable effort expended in planning a route for the Trans Canada Trail through southern BC. Serious consideration is being given to include the KVR right-of-way through the Coquihalla and Coldwater valleys from Hope to Brodie and on to Brookmere and Princeton. However, much of the route from Hope to Brodie has been alienated by pipeline and highway construction and alternatives will have to been found.

While this all may be of little consolation to current readers, we plan keep up-to-date on both subjects and post any information we can gather between editions on our World Wide Web site. Look for more information at <http://www.sonotek.com>.

•••

GPS References

We have included Global Positioning System (GPS) latitude and longitude references for many points of interest in this book. The information has been taken directly from field data files generated by a Garmin GPS 45 and/or a Garmin GPS II Plus. In general, these references should be accurate to 15 metres. However, due to the US military "Selective Availability" coding, errors of up to 100 metres are possible. Tree cover, satelite configuration and equipment accuracy may also contribute to differences in readings. The GPS references should help in locating appropriate backroad and trail junctions etc., but should not be relied on to replace maps, a compass and good "bush" sense.

•••

Emergency Numbers

At the time of writing, the 911 service was NOT available in Merritt. The following emergency numbers were in effect and should be considered up-to-date as of November, 1997. Note, that the area code is 250.

	Police	Fire	Ambulance
Douglas Lake	378-4262	n/a	1-800-461-9911
Logan Lake	523-6222	523-6262	1-800-461-9911
Merritt	378-4262	378-2211	1-800-461-9911
Spences Bridge	458-2233	458-2494	1-800-461-9911

Checklist for Camping

Back country camping can be a wonderful experience. However, it is often too easy to forget essential pieces of equipment until you are in the wilderness — then it is just too embarrassing to admit that the reason you are rubbing two sticks together is that you forgot the matches.

The following list of frequently used camping items (in alphabetical order) was provided by Jim Reid of the Coleman Camping Company. Use it when planning your next trip, then adapt it to your own needs.

[ ] Air mattress
[ ] Batteries
[ ] Blankets
[ ] Camera & Film
[ ] Coffee pot
[ ] Compass
[ ] Cooking Utensils
[ ] Cooler
[ ] Dishpan & Pot Scrubbers
[ ] Eating Utensils
[ ] First Aid Kit
[ ] Flares/Mirror
[ ] Other emergency devices
[ ] Flashlight
[ ] Folding Chairs or Camp Stools
[ ] Folding Cooler or Stove Stands
[ ] Fuel
[ ] Ground Cloth
[ ] Hand Ax
[ ] Ice or Ice Substitutes
[ ] Insect Repellent
[ ] Jug of Water
[ ] Knife
[ ] Lantern
[ ] Lighter - Disposable Butane
[ ] Mantles
[ ] Maps
[ ] Matches & Waterproof Holder
[ ] Pad, Pen or Pencil
[ ] Plastic Zipper Bags
[ ] Prescription Medicine
[ ] Radiant Heater (cold weather)
[ ] Rope, Cord or Wire
[ ] Shovel - Small Folding Type
[ ] Sleeping Bags
[ ] Snakebite Kit
[ ] Soap - Biodegradable
[ ] Stakes
[ ] Stove
[ ] Sunglasses
[ ] Suntan Oil or Lotion
[ ] Tablecloth
[ ] Tent & Poles
[ ] Toilet Paper
[ ] Toiletries
[ ] Tool Kit, Hammer & Nails
[ ] Towels - Paper & Bath
[ ] Trash Bags
[ ] Water Container
[ ] Water Purification Tablets

You may want to keep track of those pieces of equipment which you had and didn't need or needed and didn't have. This would help you on your future trips. Good camping.

•••

NEW

Coquihalla Highway 5 Exits

170 Trans-Canada Highway (Hope)
171 Third St. (Hope)
173 Thacker Rd (Hope)
177 Crowsnest Highway 3
183 Kawkawa Lake Rd
192 Sowaqua Creek
195 Carolin Mines
200 Shylock Rd
202 Portia
217 Zopkios View Point
221 Falls Lake
225 Toll Booth
228 Coquihalla Lakes
231 Mine Creek Rd
238 Juliet Creek
250 Larson Hill
256 Kingsvale (Coldwater Rd)
276 Comstock Rd
286 Route 5A (Info Centre)
290 Route 5A (Merritt-Quilchena)
315 Helmer Rd
336 Walloper Lake
355 Inks Lake Rd
362 Trans-Canada Highway (Cache Creek Exit)
366 Copperhead Dr. (Kamloops)
367 Pacific Way (Kamloops)
368 Hillside Way (Kamloops)
369 Columbia St. (Kamloops)
370 Summit Drive (Kamloops)
374 Yellowhead Highway 5 North (Jasper Exit)

Symbols used in this book

 Photography Opportunities.

 Wildlife Viewing.

 Birdwatching.

 Wildflowers.

 Interpretive Trail.

 Point of Interest.

 Viewpoint.

 Picnic Site.

 Open Picnic Shelter.

 Enclosed Picnic Shelter.

 Sleeping Shelter.

 Wilderness Campsite.

 RV Park.

 Campground.

 Campfire Ring.

 RV Sani-Station.

 Public Washrooms.

 Amphitheatre.

 Accommodations.

 Meal Service.

Symbols used in this book, continued.

Fuel.

Public Telephone.

First Aid Station.

Hospital.

Museum.

Airport.

Information Centre.

Marina.

Swimming Beach.

Angling.

Boat Launch.

Canoeing & Kayaking.

Sailing.

Motorboating.

Windsurfing.

Hiking Trail.

Rock Climbing Route.

Golf Course.

Rockhounding.

Shooting Sports.

Horse Riding.

Bicycle Riding.

Mountain Bicycle Trail.

Motorcycle Trail.

Four-Wheel-Drive Road.

Canoe Portage.

Winter Sports.

Ice Fishing.

Alpine Ski Hill.

Nordic Ski Trail.

Snowshoe Trail.

Sledding Hill.

Ice Skating.

Snowmobile Trail.

Chairlift.

Gondola Lift.

Cave.

Tunnel.

1

Hope to Merritt

Coquihalla Highway 5

Statistics	**For maps, see pages 4 and 23.**
Distance:	120 kilometres, Hope to Merritt.
Travel Time:	One to two hours.
Condition:	Paved four to six lanes. Toll Highway.
Season:	Open year around.
Topo Maps:	Tulameen 92 H/NE (1:100,000). Merritt 92I/SE (1:100,000). Yale 92H/NW (1:100,000). Chilliwack Lake 92H/SW (1:100,000).
Forest Maps:	Merritt Forest District Recreation Map. Chilliwack Forest District Recreation Map.
Communities:	Merritt & Hope.

While a freeway, or in this instance, a toll highway, is not usually considered a prime site for outdoor recreation, Coquihalla Highway 5 might be considered an exception. The route starts off at an elevation of about 50 metres near Hope, a city at the junction of the Coquihalla and Fraser Rivers, and soon begins a steady climb to the 1244 metre summit in the heart of the Cascade Mountains. From the Coquihalla summit, near the toll plaza, the divided highway follows the Coldwater River north on an undulating descent to the grasslands surrounding Merritt — at an elevation of about 600 metres.

As you will see in the following chapters, this relatively new highway through the Cascades provides easy public access to diverse terrain and a wealth of opportunities to escape the confines of civilization. As with any escape, there is a certain air of excitement and a certain amount of risk. Here the excitement of getting away must be tempered with good planning, for Nature can be harsh and unforgiving.

The Coquihalla and Coldwater valleys have no significant settlements, yet they have been in almost continuous use as a transportation corridor since the mid-nineteenth century. In 1848-49, 20 kilometres of the Coldwater valley near Merritt was part of a Hudson's Bay Company (HBC) Brigade Trail from Kamloops to the Fraser River near Spuzzum. From 1849 to the early 1860s, the Coquihalla valley from Hope to Peers Creek received steady use as part of the Hope-Tulameen section of the HBC Brigade Trail.

The Coquihalla valley saw little traffic after the completion of the Cariboo Wagon Road through the Fraser Canyon in the early 1860s. Traffic again picked up when a cattle trail was completed through the Coquihalla in 1876 at the request of Nicola Valley ranchers. The completion of the Canadian Pacific Railway (CPR) in 1885 again left the Coquihalla in disuse except for a short burst of activity when the flood of 1898 wiped out the CPR tracks in the Fraser Canyon.

The search for a route for a Coast-to-Kootenay railway led to completion of the Kettle Valley Railway (KVR) through the Coquihalla Canyon in 1916. Heavy snowfall plagued the railroaders and despite some sixteen snowsheds, slides constantly threatened the crews, sometimes trapping them for a week or more. A mudslide wiped out a major bridge in 1959, permanently closing the line. Pipeline construction, begun before the rail line closed, then turned the valley into primarily an oil and natural gas transportation corridor.

Lobbying by local residents for a new highway from Hope to Merritt began in earnest with the first Coquihalla Caravan on 21 July, 1963. After facing the indignities of the extremely rough road over the old railbed, the conquering heroes were greeted at Hope by speeches, a pipe band and all the hoopla befitting a royal entourage. For the next decade, the Coquihalla Caravan was a major summer event, attracting participants from Kamloops and the Okanagan.

The lobbying continued until 13 January, 1977 when, much to everyone's surprise, the provincial legislature announced "the design and construction of a highway from Merritt to Hope through the Coquihalla Pass." Two years later, in January 1979, Emil Anderson Construction of Hope was awarded a $4.3 million contract for the first five kilometres of the Coquihalla Freeway. Construction progressed slowly until the provincial government decided that this route to the B.C. Interior would be part of the showcase that was to become known as Expo 86. Crews began a 24 hour-a-day effort to meet the Expo deadline. Surveyors and designers barely kept ahead of earth-moving equipment. Everyone lost sight of the original budget, but the new highway opened in May 1986, in time to serve the Expo 86 tourist traffic.

Unfortunately, the term "freeway" does not really apply to the Hope to Merritt section of the Coquihalla Highway. In an effort to make the vast expenditure more politically acceptable the route became British Columbia's first toll highway in modern history. Although tolls were once in effect on the Lions Gate and Second Narrows bridges in Vancouver, and the Agassiz-Harrison Bridge across the Fraser River, toll highways were discontinued when the Cariboo Wagon Road was finally paid off more than a century ago.

Highway exit numbers will serve as references to the various recreational attractions along the route. Where justified, the major points of interest will be dealt with as separate chapters in this book.

Exit 170 (based on the number of kilometres from the Horseshoe Bay ferry Terminal in West Vancouver) is the junction between Highway 3 (The Crowsnest Route) and Highway 1 (The Trans-Canada Highway). It is the first major entrance to Hope and the point at which fuel and supplies should be checked as there are no services available between Hope and Merritt.

Exit 177 marks the junction with Highway 3 and the beginning of Coquihalla Highway 5. To the east lies Manning Park, Princeton and the Okanagan Valley. To the north is the Coquihalla Highway through the Cascade Mountains, Merritt, Kamloops and points north and east.

Once over the hump that separates Nicolum Creek from the Coquihalla River, Highway 5 follows much the same route as the fur brigades, cattle drovers and Kettle Valley Railway. For the first few kilometres, that is, for Exit 183 (Kawkawa Lake Road) marks a parting of the ways. The old HBC Brigade Trail continued east up Peers Creek and over Manson Ridge into the Sowaqua Creek drainage. (See *Peers Creek Road* on page 21.)

Five kilometres to the west of the intersection at Exit 183 the Othello (Quintette) Tunnels, built to traverse one of the narrowest constrictions in the Coquihalla Canyon, have been preserved as a provincial park. Barrie Sanford, in his book *McCulloch's Wonder*, details the almost miraculous efforts needed by Andrew McCulloch and the construction crews of the Kettle Valley Railway.

"Accepting this problem as merely another personal challenge, McCulloch daringly lowered himself and several of his fellow engineers in a small woven basket, suspended by ropes, down from the cliff tops into the canyon. There they cut narrow footholds in the rock walls upon which they could set up their survey instruments. After several weeks of such perilous work McCulloch confirmed his earlier suspicions. Instead of a single tunnel nearly a mile in length, McCulloch found that a line requiring barely a quarter as much tunneling could be cut directly

Figure 1
Othello (Quintette) Tunnels, approximatley seven kilometres east of Hope.
(Photo 413-F-33)

through the canyon. McCulloch, moreover, chose an alignment which permitted the railway to cut not one, but five tunnels, thus exposing a multiple of working faces which greatly reduced the drilling time. Two bridges were to be built between three of the tunnels, each of them flush with the adjacent tunnel portals as a result of the abrupt drop to the river abyss below. More amazingly, all five tunnels and the two bridges were set on perfect tangent alignment. By accomplishing this alignment, McCulloch was able to push the tunnels — later known as the Quintette Tunnels — through the entire canyon with a scant third of a mile of trackage, literally *'threading the canyon... (of this) awe-inspiring gorge'.*

"On July 16, 1915 the Quintette bridges were finished and the track-layers passed through the last of the five tunnels..."

Steel bridges and tracks were removed in the early 1960s after the railway discontinued operation through the Coquihalla. Bridging was restored by the Hope and District Chamber of Commerce in 1985-86, with the help of military engineers from Canadian Forces Base Chilliwack. Since then, B.C. Parks has built parking facilities and toilets a short distance north of the first tunnel.

An aerial tram, a few kilometres to the north, is used to transport technical staff to the communications sites on Jarvis Mountain.

Shylock Road (Exit 200) marks the point where the old railway right-of-way and the new Coquihalla Highway part company. The highway at this point is laid exactly where the tracks once carried steam-powered locomotives on their Coast-to-Kootenay run. The locomotives, however, continued to chug up the Coquihalla Canyon, clinging to the mountainside where no highway engineer dared build an all-weather high speed route. The highway instead takes a steeper course up the broader U-shaped Boston Bar Creek valley. Although the eight per cent grade would have been impossible for railway engines, it is only a slight inconvenience for today's traveller. The major advantages to this route are more room to build a multi-lane highway and fewer major avalanche zones.

A second aerial tramway, approximately seven kilometres north of the Shylock exit was used for avalanche control on the west side of the valley. About one and a half kilometres further north on the east side of the highway is a beautiful waterfall that is usually only visible in spring and early summer. By late summer it is a mere trickle.

If you look carefully on the west side of the road just a bit further up hill, you should see a striking box canyon. We stopped to explore it one day and discovered a black bear quietly eating the berries in the base of the canyon.

After passing through the Great Bear Snow Shed and scaling what always seems to me to be the steepest section of the road, you can detour into the Zopkios Ridge rest area (Exit 217).

In addition to year-around washroom facilities on the north side of the highway, this can be a base for some serious hiking or rock climbing. Needle Peak, to the south, can be a challenging hike. Park on the south side of the vehicle underpass and cross the creek near a small waterfall beside the highway maintenance sheds. Locate the old pipeline road and follow it downhill for about 100 metres to the trailhead. The narrow trail climbs steeply through the trees to an alpine ridge. Follow the ridge to approximately 1850 metres (6000 ft.) The final ascent to the Needle should be attempted by serious hikers only — for details, see page 25. The view is excellent even if you don't make it to the top.

While the near-white face of Zopkios Ridge can be easily photographed from the rest stop, reports are that it is easier to climb to the ridge from the Falls Lake area.

Falls Lake (Exit 221) is a destination that is attracting hikers and fishermen, particularly since the hiking trail from the end of the road to the lake was completed in 1990. (See *Falls Lake Trail*, page 29.)

The Dry Gulch Bridge, one kilometre north of the Falls Lake exit, is the tallest of the 32 river, road or rail crossings in the 120 kilometre-long-highway. A 175 metre steel arch spans the 90 metre deep gash in the mountain side.

The toll plaza, three kilometres farther north, marks the gateway to the Nicola Valley. The nearby Coquihalla Lakes drain south into the Coquihalla River, but the Coldwater River, which starts in the mountains to the west, drains north, joining the Nicola River at Merritt.

The Coquihalla Lakes exit (Exit 228) marks the departure point for several side trips and some excellent hiking and backpacking opportunities, each of which is dealt with separately. To the west is a forest road into the heart of the Cascades (See *Upper Coldwater Road*, page 33.) The road ends near the base of some challenging rock faces, but there is also an excellent hiking trail into Little Douglas Lake, near the seven kilometre point, that even the junior hiker can enjoy. (See *Little Douglas Lake Trail*, page 39.) To the east of the interchange, forest roads lead into the headwaters of the Tulameen River and a variety of Forest Service recreation sites as well as a trail along a portion of the old Hudson's Bay Company Brigade Trail. (See *Tulameen River Road*, page 45, and *Mount Davis Brigade Trail*, page 49.) Also to the east, the gravel backroads provide a shortcut to Tulameen with more Forest Service roads and recreation sites for diversion. (See *Coquihalla-Tulameen Road*, page 41.)

Figure 2
Coquihalla Highway 5, north of Comstock Road. (Photo 380-C-14)

Exit 228 also marks the return to the old railway right-of-way. A steel bridge across the Coldwater River near the interchange was once used by the KVR and CPR trains.

The Juliet Creek exit (not officially numbered), 10 kilometres north of the Coquihalla Lakes exit, is another point of departure from the paved highway. Hikers looking for a day-long jaunt can follow the logging roads up to the west into a basin near the base of July Mountain. From the end of the logging clearing, a new trail is being constructed to a beautiful alpine tarn at the foot of the north face of July Mountain. (See the *July Mountain Trails* section, page 55.)

The serious backroads explorer and fisherman can consider the forest road into Murray Lake, five kilometres to the northwest. Access from the Juliet Creek exit is steep and rough, so this backroad has been described from an easier access at Kingsvale on the Coldwater Road (Exit 256). (See *Murray Lake Road*, page 73.)

At the southern foot of Larson Hill (Exit 250) is an old railway bridge that marks Brodie, the junction of the KVR connection between Hope, Princeton and Merritt. The province purchased the right-of-way and cyclists pedal the route to Brookmere and Princeton in the summer.

The Coldwater Road also provides access to Brookmere, a tiny community once a maintenance point on the KVR, and a backroad to the alpine ridges of Thynne Mountain, both of which are covered later. Cross-country skiing trails and fine fishing lakes are accessible via Coldwater Road and the Kane Valley Road, north of Kingsvale, as well as from Highway 5A/97C, south of Merritt.

Comstock Road (Exit 276) provides access to another fishing lake and the wooded peak of Iron Mountain. (See *Gwen Lake Road* on page 93.) Beyond Comstock Road, it is another 10 kilometres to the Nicola Valley and Merritt. With the completion of the Okanagan Connector (Highway 97C) in 1990, Exit 286 became a major junction in the Coquihalla system. This exit also provides access to Merritt and points west via Highway 8. A second interchange, Exit 290, provides access to Nicola Lake and the old Kamloops-Merritt route via Highway 5A.

For a detailed look at attractions along the various routes, see the appropriate sections.

•••

GPS References for major points of interest. See page 9 for details.

Wpt	Description	Latitude	Longitude
W01	Exit 170 Highway 1 Overpass	N49° 22' 23.364"	W121° 26' 24.723"
W02	Exit 173 To Hope	N49° 22' 12.714"	W121° 24' 21.630"
W03	Exit 177 Highway 3	N49° 21' 40.825"	W121° 21' 49.354"
W04	Exit 183 Kawkawa Lake Rd	N49° 23' 17.023"	W121° 19' 10.289"
W05	Jarvis Tram	N49° 24' 28.368"	W121° 19' 03.833"
W06	Exit 192 Sowaqua Cr.	N49° 26' 52.969"	W121° 16' 13.039"
W07	Exit 195 Carolin Rd	N49° 28' 43.151"	W121° 15' 13.544"
W08	Exit 200 Shylock	N49° 30' 24.756"	W121° 11' 58.170"
W09	Waterfall	N49° 33' 58.685"	W121° 11' 04.448"
W10	Exit 217 Zopkios	N49° 35' 43.999"	W121° 07' 09.374"
W11	Coquihalla Summit	N49° 35' 40.452"	W121° 04' 58.138"
W12	Exit 221 Falls Lake	N49° 36' 41.509"	W121° 03' 04.557"
W13	Toll Booth	N49° 37' 52.919"	W121° 00' 53.326"
W14	Exit 228 Coquihalla Lakes	N49° 39' 23.541"	W121° 00' 25.745"
W15	Exit 231 Mine Cr.	N49° 41' 04.505"	W121° 00' 47.871"
W16	Juliet Cr.	N49° 44' 25.846"	W121° 00' 46.746"
W17	Exit 250 Larson Hill	N49° 49' 51.497"	W120° 56' 03.823"
W18	Exit 256 Kingsvale	N49° 52' 54.359"	W120° 54' 11.781"
W19	500 kV Lines	N49° 55' 45.192"	W120° 54' 32.260"
W20	Exit 276 Comstock Rd	N50° 01' 14.221"	W120° 49' 13.307"
W21	Exit 286 Highway 5A	N50° 05' 25.844"	W120° 45' 06.132"

•••

2

Peers Creek Road

Highway 5 to Manson Ridge HBC Trail

Statistics	**For map, see page 23.**
Distance:	Approximately 7.5 kilometres to trailhead.
Travel Time:	One half hour.
Condition:	Rough road, dry weather only.
Season:	Summer and early fall.
Topo Maps:	Chilliwack Lake 92 H/SW (1:100,000). Hope 92 H/6 (1:50,000).
Forest Maps:	Chilliwack Forest District Recreation Map.
Communities:	Merritt & Hope.

Exit 183 (Kawkawa Lake), about 10 kilometres north of Hope, provides access to one of the last remaining sections of the historic HBC Brigade Trail through the Cascade Mountains. This trail, from Fort Hope to Kamloops, was the main route of commerce to the Interior from 1849 until 1860. The wagon road built through the Fraser Canyon during Fraser River gold rush took traffic off the HBC trail.

Bob Harris and Harley Hatfield, in *Old Pack Trails in the Cascade Wilderness*, write that Alexander Caulfield Anderson made the first recorded trip through the Cascades in 1846. The HBC wanted a route from Kamloops to the Pacific to replace the trail to the Columbia River.

Anderson left the Fraser River at what is now Hope and followed much the same route as Highway 3 to Snass Creek. He followed Snass Creek upstream, continuing north up the east fork to the Cascade divide. After reaching the "Punch Bowl" near the headwaters of the Tulameen, Anderson followed the river down through its narrow canyons to Otter Lake. Near the present village of Tulameen, Anderson met the Native chief "Blackeyes" who told him that, but for the snow, he would have seen the easier trail running directly over the Tulameen Plateau.

Based on his initial experience, Anderson later recommended a trail from Yale to the Nicola Valley via the Anderson River, Uztlius Creek,

Maka Creek, Spius Creek, Midday Creek and the Coldwater River. The Kequeloose Trail, as it was known, proved unsatisfactory because of the dangerous Fraser River crossing and little feed for the pack horses.

Figure 3
Descending Manson Ridge to Peers Creek — August 1980. (Photo 270-F-34)

Henry Peers was assigned the task of finding a new route through the mountain barrier and he chose to explore the second major Coquihalla tributary northeast of Hope. Travelling upstream near the creek that now bears his name, he crossed Manson Ridge and descended into the Sowaqua basin via Fool's Pass. A sloping, wooded ridge took him to the valley floor near the mouth of Matthew Creek. Peers' trail then climbed Mount Davis and descended into the headwaters of Podunk Creek, a tributary of the Tulameen River. (See *Mount Davis Brigade Trail*, page 49.) After reaching the Tulameen, he followed Blackeyes' trail over the top of the Tulameen Plateau and down to Otter Lake. From "Campement des Femmes", near present-day Tulameen, the overland route to Kamloops was relatively easy.

Peers' trail was the main route of commerce from 1849 to 1860. Then the discovery of gold in the upper reaches of the Fraser River prompted the construction of a wagon road through the nearly impassable Fraser Canyon. New railway and highway routes replaced the HBC Brigade Trail and it had almost disappeared when, in the late 1970s, **Opportunities For Youth** work crews began rebuilding it under the auspices of the Okanagan Similkameen Parks Society.

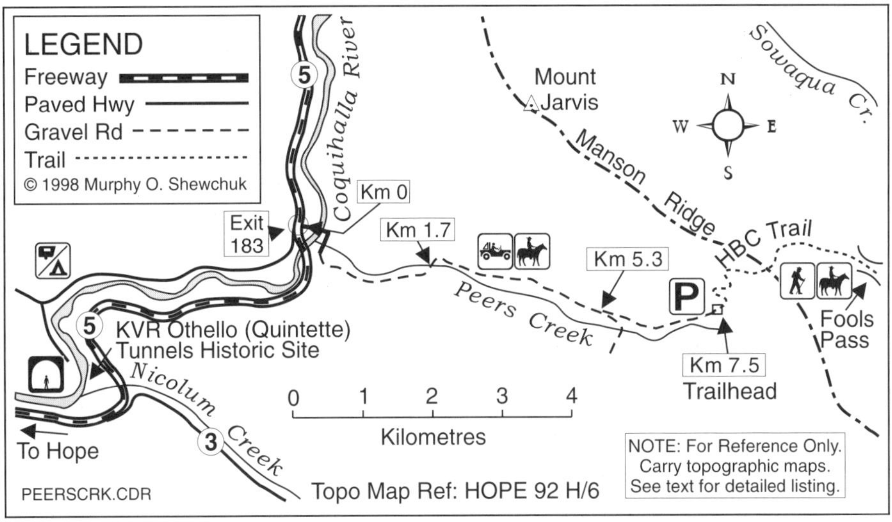

Map 2
Othello (Quintette) Tunnels — Peers Creek area.

As late as the mid-1980s, the Brigade Trail from Lodestone Lake to the head of Peers Creek was total wilderness. Logging roads have since been built across Podunk Creek and along the Sowaqua. What was once a week-long hike can now be divided into several sections, although wilderness hiking experience is still essential.

To reach the start of the trail up Manson Ridge, take Exit 183 (Kawkawa Lake Road) and cross over to the east side of the Coquihalla River. With the interchange as your km 0 reference, follow the Peers Creek Forest Road upstream, crossing the bridge to the north side of the creek near km 1.7. The rough, narrow road climbs steeply for the next 2.5 kilometres, gaining almost 500 metres in altitude. Keep right at a junction near km 4.4 and left at a junction near km 5.3. The road to the right at km 5.3 swings around the head of the valley. The side road to the left climbs steadily to the Manson Ridge trailhead near km 7.5.

From the end of the old road, the elevation gain to the ridge is about 500 metres. At the time of writing, there were no signs at the start of the trail. It climbs north through the trees just east of the old cut block. It is marked with red squares tacked to the trees and small rectangular metal markers designating it as the HBC Brigade Trail.

The journal of Lt. H.S. Palmer, RE describes his climb up 1450 metre high Manson Ridge from Manson Camp (near the present trailhead) in mid-September 1859: "Rose at dawn, and soon commenced the laborious ascent of the Mountain by a zigzag trail... After struggling up this difficult mountain path for an hour and a half, we reached the summit of the pass, the magnificent view from which fully compensates...for the labour.... As far as the eye could reach, an endless sea of mountains rolled away into blue distance... How we were ever to get out of them appeared to me somewhat difficult of solution..."

Lt. Palmer's views were not always as beautiful, for he also reports: "Mr. McLean of the Hudson Bay Company, who crossed in 1857 or 1858, on the 16th of October, had a very disastrous trip, and lost 60 or 70 horses in the snow...in riding over the mountain, and more particularly on its eastern slope, my horse frequently shied at the whitened bones of some one of the poor animals...left to perish where he lay."

The trail breasts the ridge near the west end of a long natural clearing. Various game trails follow the ridge to the east, providing access to alpine meadows and some excellent views of the lower Fraser Valley and the Cascade Mountains to the north. The trail down to Fool's Pass and Sowaqua Creek may be difficult to locate in places because of logging and new growth.

NOTE: In August 1997, Peers Creek Road was passable for about six kilometres with a vehicle with good clearance and hill-climbing ability. Logging preparation was underway. Beware of logging trucks.

•••

GPS References for major points of interest. See page 9 for details.

Wpt	Description	Latitude	Longitude
W01	Exit 183 (Kawkawa Lake Road)	N49° 23' 07.367"	W121° 19' 01.067"
W02	Second Bridge across Peers Cr.	N49° 22' 48.934"	W121° 18' 12.480"
W03	HBC Manson Ridge Trailhead	N49° 22' 21.205"	W121° 13' 45.386"

•••

3

Needle Peak Trail

Statistics	For map, see page 28.
Distance:	Approximately 5.5 km, Rest Area to Needle Peak.
Travel Time:	Approximately 6 to 7 hours round-trip hiking time.
Elevation Gain:	Approximately 885 metres.
Condition:	Strenuous hiking trail. Lower section can be slippery.
Season:	July through September.
Topo Maps:	Spuzzum 92 H/11 (1:50,000).
Forest Maps:	Chilliwack Forest District Recreation Map.
Communities:	Merritt & Hope.

Needle Peak is the prominent pyramid-shaped peak that stands above the rest of the mountains between the Coquihalla Canyon and Boston Bar Creek. The original access was from the KVR or the road through the bottom of the Coquihalla Canyon. However, the new highway route through the Boston Bar Creek valley saves at least 300 metres of climbing and bushwacking.

The trail is not well marked, but once you find it, it should be easy to follow. It starts from the old pipeline access road on the south side of Exit 217 at the Zopkios Ridge rest area. Park on the north side of the highway or well out of the way of the highway maintenance equipment and buildings on the south side. Cross the creek to the right (west) of the buildings — the boulders near the base of a small waterfall may prove to be good stepping stones. Then follow the old road downhill for about 50 to 100 metres, watching for survey ribbons or metal markers on the trees on the uphill side of the road.

The trail starts off at an elevation of about 1220 metres. As it climbs through the thick timber, it gets progressively steeper. At about 1550 metres, it leaves the trees as it reaches the crest of a ridge that extends to the west shoulder of Needle Peak. Once out of the timber, the climb is much gentler. You'll have your choice of routes to the 2103 metre peak, but one that swings around the back (south) side of the right shoulder may prove easier.

I suggest strong hiking boots with good tread as the trail through the timber can be slippery. You should carry plenty of water, particularly if you are travelling in late August — a good time to try this trail.

•••

Figure 4
Reaching the ridge on the Needle Peak Trail. (Photo NED-002)

GPS References for major points of interest. See page 9 for details.

Wpt	Description	Latitude	Longitude
W01	Highway Maintenance Yard	N49° 35' 43.656"	W121° 07' 24.184"
W02	Start of trail	N49° 35' 41.936"	W121° 07' 33.965"
W03	Break out of timber	N49° 35' 12.360"	W121° 08' 05.383"
W04	Cairn on ridge	N49° 34' 52.202"	W121° 08' 08.492"
W05	Head of ridge	N49° 34' 21.829"	W121° 08' 06.662"
W06	Needle Peak	N49° 34' 13.270"	W121° 07' 27.999"

•••

Figure 5
Needle Peak from near the crest of the ridge.

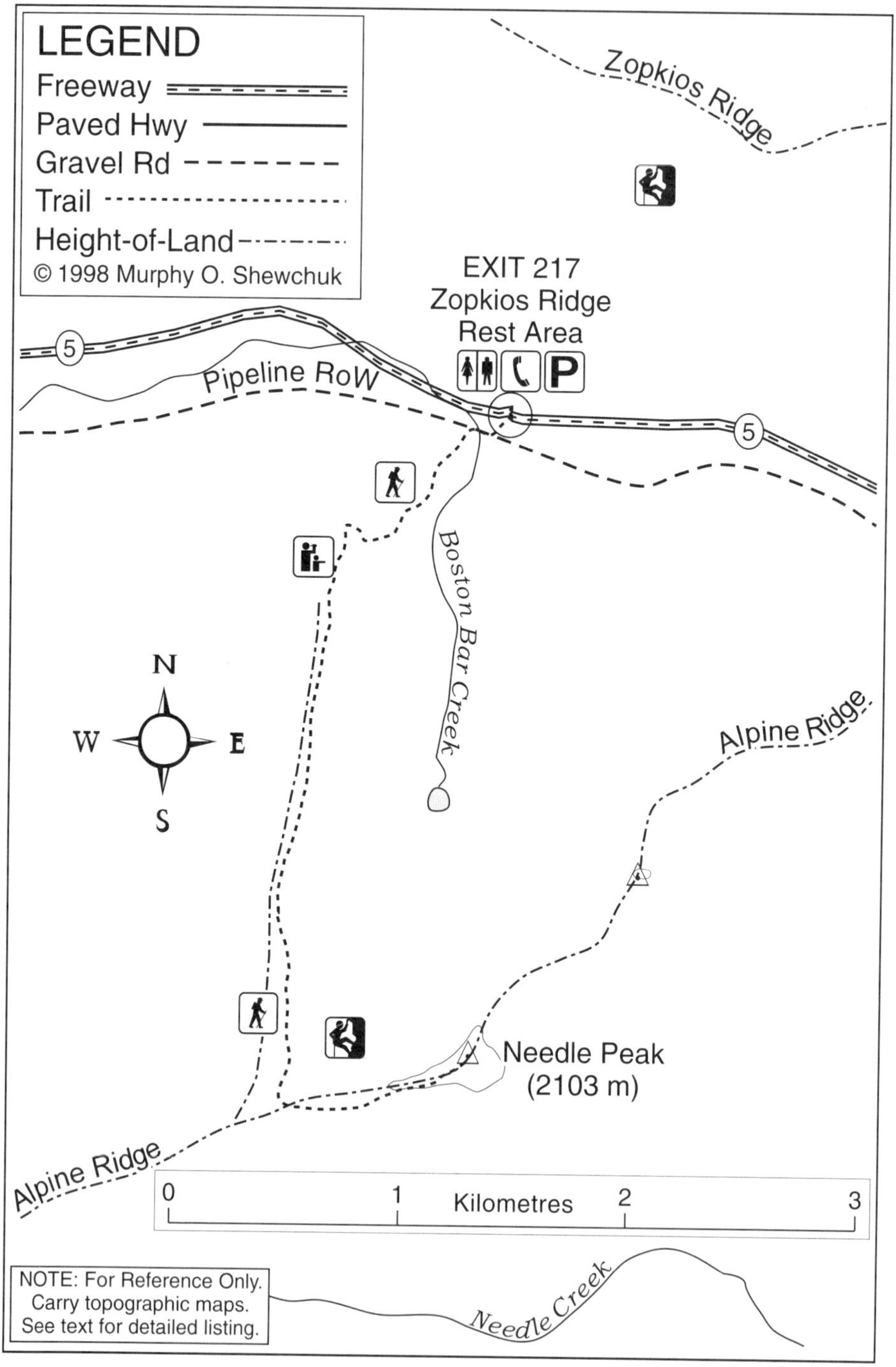

Map 3
Needle Peak Trail

4

Falls Lake Trail

Statistics	**For map, see page 30.**
Distance:	1.1 km, Coquihalla Highway 5 to trailhead. Approximately 1 kilometre trailhead to Falls Lake.
Travel Time:	Approximately 20 minutes hiking time.
Condition:	Good walking trail.
Season:	All season. Beware of avalanches in spring.
Topo Maps:	Spuzzum 92 H/11 (1:50,000).
Forest Maps:	Chilliwack Forest District Recreation Map.
Communities:	Merritt & Hope.

Falls Lake takes its name from beautiful Bridal Veil Falls, a famous KVR landmark in the Coquihalla Canyon. One of the last remaining railway bridges straddles Fallslake Creek. The bridge can be seen from the rough gravel road through the canyon.

Take Exit 221 and drive west for about one kilometre. The gravel road to the parking area can be rough, however it is solid. After parking, cross the creek and follow the trail westward to Falls Lake. The trail climbs an old skid road, then enters what resembles a coastal rain forest. Boardwalk bridges traverse the small creeks and log-reinforced steps make the rough terrain easy to navigate.

The elevation gain is minor. The parking lot is about 1250 metres and the lake is about 1300 metres with the trail climbing little more than 10 metres above lake level as it skirts the cliffs. Trout fishing, the easy walk and the cool mountain air are the prime attractions of Falls Lake.

•••

GPS References for major points of interest. See page 9 for details.

Wpt	Description	Latitude	Longitude
W01	Exit 221 Falls Lake	N49° 36' 37.501"	W121° 03' 12.466"
W02	Falls Lake Trail Head	N49° 36' 43.604"	W121° 03' 53.832"
W03	Falls Lake Outlet	N49° 36' 55.586"	W121° 04' 22.015"
W04	Viewpoint at top of Dry Gulch	N49° 37' 40.149"	W121° 03' 23.397"

•••

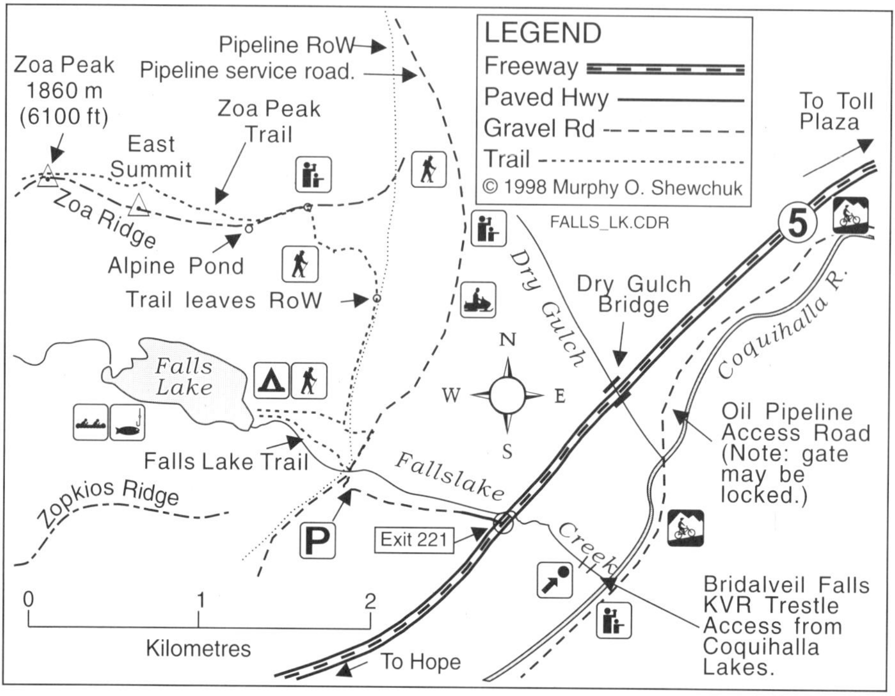

Map 4
Falls Lake — Zoa Peak area.

Figure 6
Falls Lake, looking west from the outlet. (Photo 426-A-5)

NEW

5

Zoa Peak Trail

Statistics	For map, see page 30.
Distance:	1.1 kilometres, Coquihalla Highway 5 to trailhead. Approximately 4 kilometres, trailhead to Zoa Peak.
Elevation Gain:	Approximately 610 metres.
Travel Time:	Approximately 5 hours round trip hiking time.
Condition:	Good walking trail. A few steep sections.
Season:	Summer trail. Beware of avalanches in spring.
Topo Maps:	Spuzzum 92 H/11 (1:50,000).
Forest Maps:	Chilliwack Forest District Recreation Map. Merritt Forest District Recreation Map.
Communities:	Merritt & Hope.

The trail to Zoa Peak is probably one of the least strenuous hikes to the summit of any of the "animal" peaks in this area of the North Cascade Mountains. It offers several excellent views of the more challenging peaks and the Upper Coldwater Valley to the northwest.

In early summer, the alpine and sub-alpine meadows can be a wonderland of colour. Lupines are common on the lower elevations. Red and white heather carpet the sheltered slopes nearer the summit. As the seasons progress, the dominant flora shifts to wild berries. On a mid-September hike, we found an abundance of juicy, ripe blueberries and black huckleberries all the way from the Falls Lake Trail parking lot to the 1860 metre summit of Zoa Peak.

At the time of writing, the trail was not signed. However, there were plenty of survey ribbons in the upper section of the trail. For the most part, it should not be difficult to follow. Cross the creek and hike north on the pipeline maintenance road for about 200 metres. Watch for a rough road coming in from the left, then climb this steep (now de-commissioned) road toward Falls Lake for another 150 to 200 metres. At the crest, turn right on the broad pipeline right-of-way climbing toward the north. In mid-September 1997, the brush was shoulder-high in places but the trail was still visible as it wound up the right-of-way.

After a half-hour hike (approximately 700 metres distance), the trail leaves the right-of-way and begins a meandering climb through the semi-open timber. The point of departure was marked with a row of boulders and numerous strips of survey ribbon. If, for some reason these are missing, watch for a rocky staircase up the right-of-way cut bank.

The trail through the trees is well-worn and has been recently tagged with pink survey ribbon. In several places it climbs through natural openings created by the underlying rock outcroppings. After about an hour's climbing (1.25 kilometres), you should reach several viewpoints and an excellent spot for a break.

Watch for signs of an old campfire near the trail. Near here is a small sub-alpine pond a few metres off the trail to the west. This pond could be one of your few sources of water in late summer and early fall. It will likely require filtering and purification to be safe. Because of the limited water near the trail, you should carry plenty of drinking water.

Up to this point, the trail has been generally on the south side of the ridge. Now it begins to follow the crest of the ridge, gradually swinging over to the north side near the first summit, some 600 metres later. The trail skirts the first summit, descends 20 to 30 metres in altitude while crossing a V-shaped cut, then climbs to the main (west) summit. The climb out of the cut could be slippery in wet conditions, but otherwise the trail does not appear to be particularly hazardous.

Scattered timber on the main summit means a bit of scrambling around to enjoy the various views. Little Douglas Lake lies far below to the west. You may catch a glimpse of the flat top of Mount Baker through a mountain pass to the south.

•••

GPS References for major points of interest. See page 9 for details.

Wpt	Description	Latitude	Longitude
W01	Exit 221 Falls Lake	N49° 36' 37.501"	W121° 03' 12.466"
W02	Falls Lake trail parking	N49° 36' 43.943"	W121° 03' 55.045"
W03	Jct. with service road	N49° 36' 52.182"	W121° 03' 48.090"
W04	Jct. old road & pipeline RoW	N49° 36' 54.082"	W121° 03' 55.192"
W05	Jct. pipeline RoW & trail	N49° 37' 16.140"	W121° 03' 47.738"
W06	Viewpoints	N49° 37' 30.875"	W121° 04' 22.698"
W07	Small sub-alpine pond	N49° 37' 28.291"	W121° 04' 23.098"
W08	East Zoa Peak	N49° 37' 36.722"	W121° 04' 53.026"
W09	Centre of defile	N49° 37' 38.125"	W121° 05' 05.385"
W10	Main Zoa Peak	N49° 37' 38.131"	W121° 05' 25.325"

•••

6

Upper Coldwater Road

Statistics	For map, see page 34.
Distance:	Approximately 15 kilometres to end of road.
Elevation Gain:	250 metres.
Travel Time:	One half hour.
Condition:	Gravel road, some rough sections.
Season:	Summer and early fall. Also suitable for skiing.
Topo Maps:	Spuzzum 92 H/11 (1:50,000).
Forest Maps:	Merritt Forest District Recreation Map.
Communities:	Merritt & Hope.

Take a close look at a map of the area to the west of the Coquihalla Lakes and you may be surprised to see the unusual names that adorn the craggy peaks and ridges. According to Pat Lean in *Place Names of the Nicola Valley* (Nicola Valley Archives Association Newsletter/Historical Work Paper July 1978) the group of mountains that surround the headwaters of the Coldwater and Anderson rivers received their strange names in the late 1970s.

George M. Dawson's "Preliminary Report of the Physical and Geological Features of the Southern Portion of the Interior of British Columbia — 1877", published by the Geological Survey of Canada in 1879, (pp 37-38) describes the area in this manner:

"About the junction of the Uz-tli-hoos and Anderson, the mountains have an average elevation of between 5,000 and 6,000 feet. Looking up the Valley of the Anderson, a great block of higher mountains at a distance of about thirteen miles, can be seen. On the west of the group is an irregular conical peak, nearly vertical on one side. These summits must reach an altitude of 7,000 to 8,000 feet. West of the immediate vicinity of the Coldwater, on this line, no country suitable for stock-raising or farming exists. Fine timber occurs in some of the valleys, as in all those of streams draining the eastward slopes of the Cascades. Here, however, it is almost inaccessible..."

Although Dawson does not appear to have assigned a name to these mountains, Amos Bowman may have done so in 1882-84. The name Anderson River Mountains assigned originally to the whole group seems officially to have been reduced to the one peak referred to by Dawson "an irregular conical peak, nearly vertical on one side..."

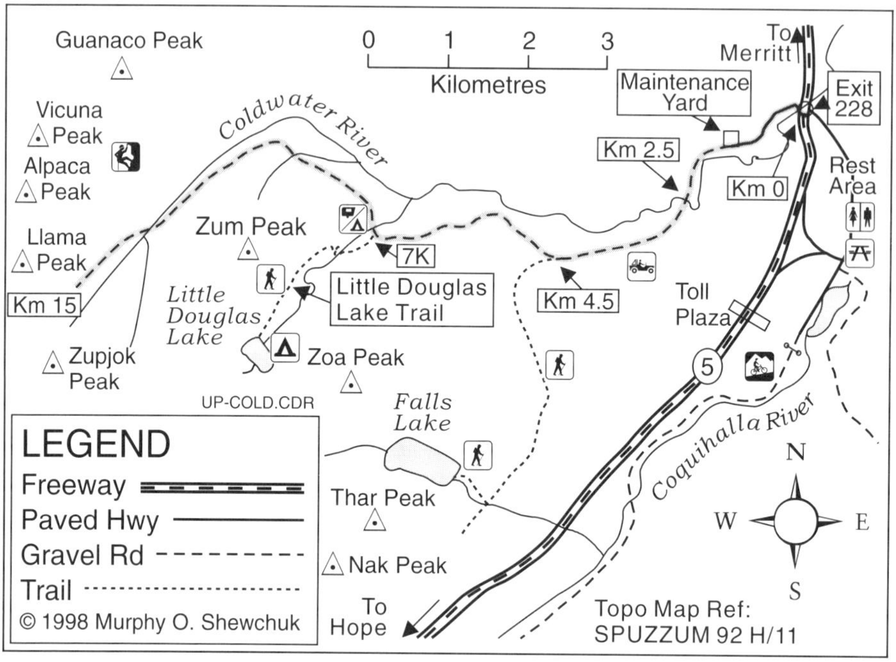

Map 5
Upper Coldwater River - Little Douglas Lake area.

However, on December 7, 1975, nearly 100 years after Dawson's visit, Philip Kubic, a mountaineer from West Vancouver, put forth a proposal to have the individual mountain peaks officially named. The request was based on the confusion caused to mountaineers by the lack of official names and included a whole series of proposed names. Kubic's proposal was accepted by the Canadian Permanent Committee on Geographical Names.

In naming these mountains, Philip Kubic chose the names of ruminants, (animals that chew their food twice) who inhabit alpine regions of the world. These he divided into three groups. European for the most westerly mountains; American for the centre group of features; and Asian for the easterly group. They follow in alphabetical order:

Alpaca Peak Between headwaters of Coldwater and Anderson Rivers. Alpacas are domesticated relatives of Vicunas and Guanacos.

Anderson River Mountain Lying to the north west of the group of mountains this is probably the one specific feature pointed out by Dawson in 1877. In writing about it in 1975 Kubic stated: "Unfortunately the name Anderson River Mountain doesn't fit into the scheme but I think that this isn't a serious drawback. All of the peaks that I have proposed names for are granitic. Anderson River Mountain, however, is composed of a type of brown rock quite different from granite. Geologically, Anderson River Mountain is unrelated to its neighbors. This peak, of course, takes its name from the Anderson River which was named in honor of Alexander Caulfield Anderson. (See *Peers Creek Road*, page 21.) It is interesting to note that this mountain was indicated on Anderson's map compiled in 1867 with the words "Conical Peak."

Bighorn Peak Near the headwaters of the East Anderson River. The name refers to the Rocky Mountain Bighorn sheep.

Chamois Peak To the east of Anderson River and just to the south of Anderson River Mountain. A Chamois is a small horned species of antelope which inhabits the Alps. This name may be appropriate in British Columbia since a chamois or "shammy," being the leather made from the skin of a Chamois, was used by early placer miners to separate the gold from the mercury amalgam as part of the refining process.

Gamuza Peak To the east of Anderson River. The most southerly of the western group of peaks. Gamuza is the Spanish work for Chamois.

Gemse Peak Second range of peaks to the east of Anderson River. Gemse is German for Chamois.

Guanaco Peak At the headwaters of the Coldwater and East Branch of Anderson River. A Guanaco is a species of Llama inhabiting the Andes from Peru to Tierra del Fuego.

Ibex Peak To the east of Anderson River and south of Chamois Peak. An Ibex is a species of goat with very long curved horns inhabiting the Alps.

Llama Peak At the headwaters of the Coldwater and Anderson Rivers. Llamas are domesticated relatives of Vicunas and Guanacos.

Nak Peak Above Zopkios Ridge, near the headwaters of the Coldwater River and to the west of the Coquihalla River. A Nak is a female Yak.

Reh Peak One of that group of eight peaks lying to the east of Anderson River and to the south and west of an unnamed branch of Anderson River. A Reh is a deer which inhabits the foothills of the Alps.

Serna Peak To the south and west of Reh Peak. Serna is Russian for Chamois.

Steinbok Peak To the south and east of Anderson River Mountain. Steinbok is the German term for Ibex.

Thar Peak The most easterly of three peaks rising above Zopkios Ridge, it lies wholly within the Coquihalla River watershed. A Thar is

Himalayan animal similar to a Chamois. It is a species of goat-like antelope inhabiting high inaccessible mountains in Europe and western Asia. It is about the size of a full-grown goat and extremely agile.

Figure 7
Vicuna Peak from the west end of the Upper Coldwater Road.

Vicuna Peak At the headwaters of the Coldwater River and East Anderson River. A Vicuna is a species of Llama inhabiting the mountains of Peru, Bolivia and Colombia. It is a valuable source of wool.

Yak Peak One of the three peaks rising above Zopkios Ridge at the head of the Coldwater River and to the west of the Coquihalla River. A Yak is a horned wild ox inhabiting Tibet and the Himalayas.

Zoa Peak Lies at the headwaters of the Coldwater River and Fallslake Creek. A Zoa is the female offspring of a bovine bull and a Nak.

Zopkios Ridge A ridge lying to the north of Nak, Yak and Thar Peaks at the headwaters of Boston Bar Creek and Fallslake Creek. A Zopkios is either the male or female offspring of a bovine bull and a Nak.

Zum Peak Lies completely within the several branches that form the headwaters of the Coldwater River. A Zum is the female offspring of a bovine bull and a Nak.

Zupjok Peak At the headwaters of Boston Bar Creek, Anderson River and Coldwater River. A Zupjok is the male offspring of a bovine bull and a Nak.

Philip Kubic is to be commended for his efforts towards having these features properly designated. His originality certainly has added color to the place names of British Columbia.

Figure 8
Huckleberries are plentiful in the upper Coldwater. (Photo BHB-043)

Exit 228, which provides access to the Tulameen area, also allows you to explore the upper reaches of the Coldwater River. Head west, across the overpass if you are coming from the south, and cross the Coldwater River on a steel-sided bridge that once served the Kettle Valley Railway. With this bridge as your reference, you will reach the end of pavement at the highway maintenance yard near km 1 and begin the gravel Forest Road that winds into the heart of the mountains.

The road crosses the Coldwater River again near km 2.5. Just downstream from the bridge, the Coldwater makes a wide swing in high water, creating an island. In low water, if you are willing to risk getting stuck in loose gravel, you can drive down the stream bed and camp on the island. Carry a shovel and be extremely careful with fire as there are no facilities on this island.

Keep to the right at a junction near km 4.5 and follow the main valley upstream through several logged areas. The Zum Peak Forest Service recreation site, at a creek crossing just past the 7K sign, marks the entrance to the hiking trail into Little Douglas Lake. (See *Little Douglas Lake Trail*, page 39.)

Upper Coldwater Road then swings to the north around Zum Peak before continuing southeast to the headwaters of the Coldwater River. A junction in the middle of the valley near the 13K sign marks a side road to a cut block on the eastern slopes of Vicuna and Guanaco peaks. At the time of writing, the road ended at an elevation of approximately 1350 metres, in a cut block about 15 kilometres from the Coquihalla Highway. Old skidder trails lead to the upper fringes of the timber, a short hike from the sheer granite slopes of Llama and Alpaca peaks.

Although you are never far from logging cuts, the huckleberries, new growth and easier access to some spectacular mountain peaks do make it a little easier to accept.

•••

GPS References for major points of interest. See page 9 for details.

Wpt	Description	Latitude	Longitude
W01	Exit 228 - Highway 5	N49° 39' 20.938"	W121° 00' 29.454"
W02	Former KVR Bridge	N49° 39' 29.956"	W121° 00' 35.094"
W03	Highway Maintenance Yard	N49° 39' 14.777"	W121° 00' 59.446"
W04	Coldwater R. Bridge	N49° 38' 48.262"	W121° 01' 49.463"
W05	Jct. to Dry Gulch	N49° 38' 31.441"	W121° 02' 33.089"
W06	Zum Peak FS Rec. Site	N49° 38' 24.528"	W121° 05' 36.726"
W07	Jct. to Vicuna Peak	N49° 38' 23.987"	W121° 08' 15.258"
W08	Coldwater R. Bridge	N49° 38' 14.717"	W121° 08' 27.810"
W09	End of Road	N49° 38' 13.597"	W121° 08' 33.990"

•••

7

Little Douglas Lake Trail

Statistics	For map, see page 34.
Distance:	Approximately 1.7 kilometres, trailhead to lake.
Elevation gain:	Approximately 100 metres.
Travel Time:	30 to 40 minutes each way.
Condition:	Well-marked forest trail.
Season:	Summer and early fall.
Topo Maps:	Spuzzum 92 H/11 (1:50,000).
Forest Maps:	Merritt Forest District Recreation Map.
Communities:	Merritt & Hope.

Little Douglas Lake is surrounded on three sides by jagged mountain ridges reaching as high as 1850 metres — a jewel in a rough-hewn setting. Although less than an hour's walk from the Upper Coldwater Road, the tall timber and high ridges surrounding the lake easily hide the scars and sounds of civilization.

The Little Douglas Lake Trail and a rudimentary campsite near the outlet of the lake were first brushed out in 1977. At that time there were also plans to continue the trail up the north ridge and along the crest of Zopkios Ridge to the outlet of Falls Lake. The second part of the plan has so far not happened, but the Forest Service has continued to improve the trail and the campsite at the lake.

The Little Douglas Lake Trail begins at Zum Peak Forest Service Recreation Site, located at a bridge on the Upper Coldwater Road, approximately 7.5 kilometres from Highway 5. The trail initially meanders through a shoulder-high jungle of fireweed and shrubs, including plenty of huckleberry bushes.

After two or three hundred metres, the trail enters the timber and crosses the stream before climbing up to a bench on the northwest side of the valley. It is then an easy walk through the forest to the lake. Hemlock, cedar, ferns and the occasional devil's club line the 1.7 kilometre trail, but it is not as overgrown as some coastal trails.

Figure 9
Little Douglas Lake, in the Upper Coldwater River area. (Photo CLD-140)

Little Douglas Lake is nearly circular with a small sheltered bay on the southeast corner. The glacial silt has built a marsh on the northeast side, but the rocky slopes on the north and south provide easier access to open water for a little trout fishing. Don't expect any lunkers as the trout seen from shore appear to be little more than 25 cm long.

•••

GPS References for major points of interest. See page 9 for details.

Wpt	Description	Latitude	Longitude
W01	Zum Peak FS Rec. Site	N49° 38' 24.528"	W121° 05' 36.726"
W02	Trail Footbridge	N49° 38' 12.303"	W121° 05' 53.122"
W03	Little Douglas Lake	N49° 37' 48.974"	W121° 06' 21.877"

•••

8

Coquihalla-Tulameen Road

Statistics	**For map, see page 43.**
Distance:	40 km from Coquihalla Highway 5 to Tulameen.
Travel Time:	One to two hours each way.
Condition:	Gravel road with a few rough, narrow sections.
Season:	Best during dry weather. May be closed in winter.
Topo Maps:	Tulameen 92 H/NE (1:100,000). Princeton 92 H/SE (1:100,000).
Forest Maps:	Merritt Forest District Recreation Map.
Communities:	Merritt, Hope & Tulameen.

There is an interesting — and sometimes challenging — backroad from Coquihalla Highway 5, just north of the Toll Plaza, to the community of Tulameen. If conditions are right, it can save you a few hours of travelling as well as give you a chance to explore some new terrain.

Road conditions in the Cascade Mountains depend on a number of factors, most important of which are season and weather. The best time to travel through this area is summer and early fall. However, sudden storms on the higher ridges can quickly turn the clay and gravel surface to mud — good all-season tires are worthwhile at all times. Also useful insurance are an ax or saw and a full tank of fuel.

Coquihalla Lakes Exit 228, a short distance north of the Toll Plaza, is the departure point for the Britton Creek Rest Area as well as several other interesting backroad trips. The road to the village of Tulameen begins as the Tulameen River Forest Service Road, a few hundred metres south of the Rest Area entrance.

With the junction near the old Kettle Valley Railway (KVR) "Y" as the km 0 reference, follow the well-used gravel road as it gradually climbs south and east of the Coquihalla Lakes. The road leaves the Coquihalla Valley near km 2. It begins a steady climb up and over a barely perceptable divide into the Britton Creek watershed as it swings eastward. A junction near 8K marks a name change for the route to Tulameen. The Tulameen River Forest Service Road continues to the right (south) to the headwaters of the Tulameen River. (See *Tulameen River Road*, page 45.) Follow the Lawless-Britton Creek Forest Road to the left as it continues a southeasterly course along Britton Creek.

Figure 10
East Murphy Lake from the Lawless-Britton Creek Forest Service Road.

A junction to the right near km 11 marks a side road down to the west end of the Murphy Lakes and a Forest Service recreation site a short hike from the lake shore.

The road continues to climb as it swings away from Britton Creek, offering an excellent view of the Murphy Lakes at km 14.5.

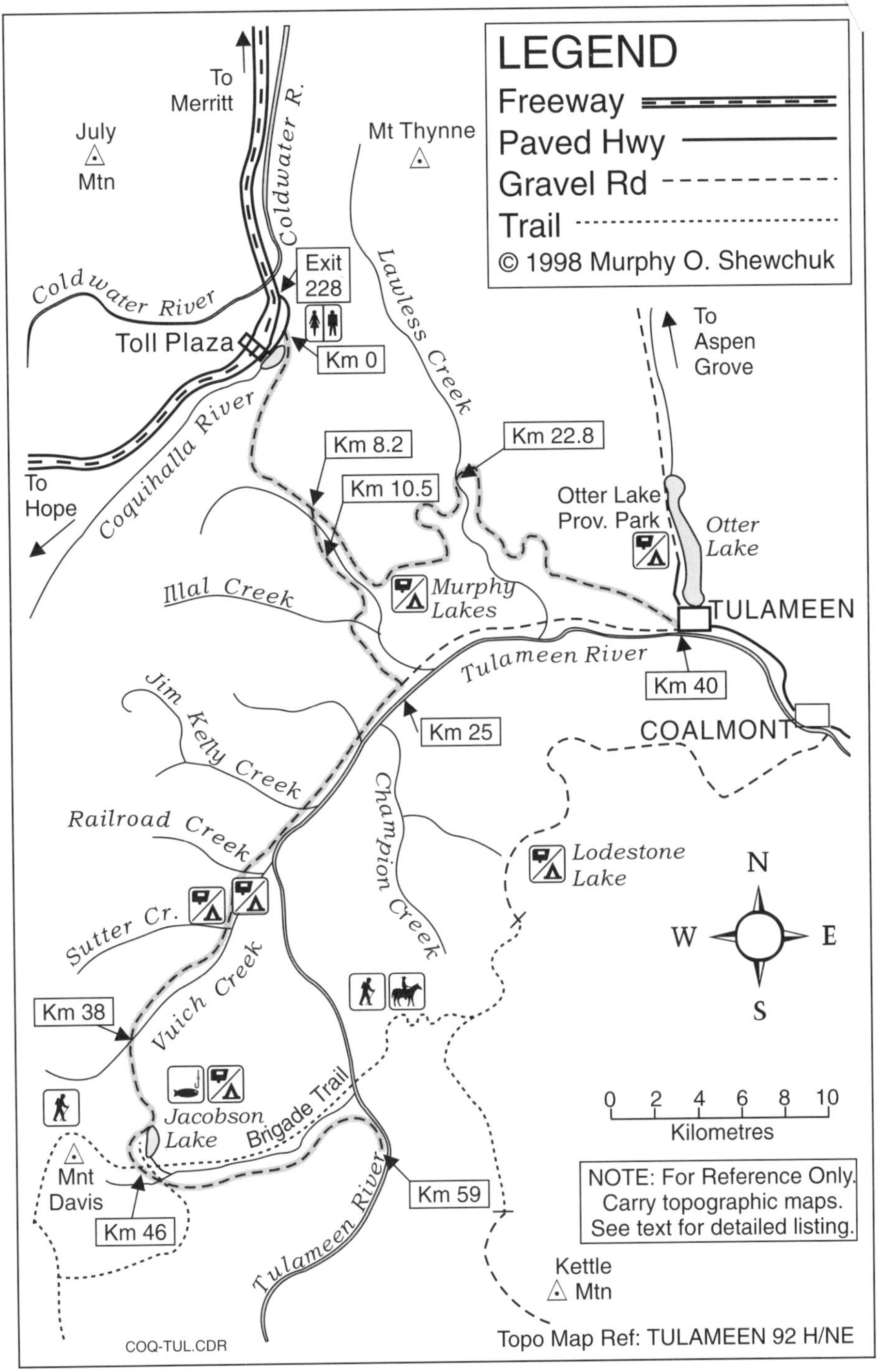

Map 6
Coquihalla Summit — Tulameen River area.

The road now swings northwest, then north before crossing Skwum Creek at km 20.3. Keep right and across the bridge unless you are looking for a dead-end road to camp or try a little hunting. If you are interested in a little exploring, Merritt Conservation Officer Dave Crack says that the road to the left, just past the Skwum Creek bridge, leads to an old mine near the top of Mount Henning.

From Skwum Creek the road traverses the hillside before descending to Lawless Creek at km 23. There are several openings near the creek for self-contained camping. Carry a shovel, for biffies are in short supply. One of these openings is the access road to a placer mining operation that was active in the late 1980s.

After crossing Lawless Creek, the road begins a switch-back climb up the side of Mount Rabbitt. A junction at a turn near km 25 marks another change in identity as the road becomes the Lawless Creek Forest Service Road. Keep right and you will soon note that the roadside kilometre markers are beginning their countdown to Tulameen. The road crests at an elevation of about 1450 metres and begins the long brake-burning descent to Tulameen. Keep to the old road past a new forest road and some ranches and you should reach Tulameen via the Lawless Creek Road about 40 kilometres after leaving the Coquihalla Lakes area.

If you are planning to take this shortcut from the east end, follow 2nd Street in Tulameen west to Lawless Creek Road, then backtrack the previous description. Double-check your kilometre readings at the junctions and carry maps as some of the roads can be confusing.

•••

GPS References for major points of interest. See page 9 for details.

Wpt	Description	Latitude	Longitude
W01	Start Tulameen River FS Road	N49° 38' 56.045"	W121° 00' 03.364"
W02	Lawless-Britton Cr. FS Road	N49° 35' 20.176"	W120° 59' 11.589"
W03	3K Junction To Murphy Lake	N49° 34' 26.412"	W120° 57' 25.373"
W04	Murphy Lk West Rec Site	N49° 33' 12.331"	W120° 55' 53.314"
W05	Skwum Creek	N49° 34' 57.040"	W120° 54' 27.492"
W06	Lawless Cr. Bridge	N49° 35' 20.910"	W120° 53' 03.658"
W07	Lawless Cr. Road	N49° 35' 59.669"	W120° 52' 41.488"
W08	Rabbitt Mtn Road	N49° 33' 19.226"	W120° 48' 31.766"
W09	Tulameen (Downtown)	N49° 32' 38.709"	W120° 45' 31.798"

•••

9

Tulameen River Road

Statistics	For maps, see pages 43, 46 and 50.
Distance:	60 km, Coquihalla Highway 5 to end of road.
Travel Time:	One to two hours in dry weather.
Condition:	Gravel road with rough sections.
Season:	Summer and early fall.
Topo Maps:	Tulameen 92 H/NE (1:100,000). Princeton 92 H/SE (1:100,000).
Forest Maps:	Merritt Forest District Recreation Map.
Communities:	Merritt & Hope.

I had mixed feelings when the Tulameen River Road was extended up Vuich Creek in the late 1980s. The Tulameen-Hope section of the historic Hudson's Bay Company (HBC) Brigade Trail was now intercepted near its mid-point. The wilderness aspect was lost, but access was improved. Obviously there are winners and losers in the sharing of natural resources. As a backroads explorer, I was a winner, for the road opened up some beautifully scenic mountain country. As a wilderness hiker, I became a loser, for now the negative aspects of "civilization" are an unwelcome part of the environment.

Logging and mining roads and the Coquihalla Highway have improved access to the Cascade Mountains, but not without opposition. The Okanagan Similkameen Parks Society (OSPS) has waged a lengthy campaign to preserve the last remaining wilderness section of the old HBC Brigade Trail. The Hudson's Bay Company used the rugged trail between 1848 and 1860 as a trade route from Hope to Kamloops, using packhorses to carry furs and trade goods.

My wife, Katharine, and I hiked the Brigade Trail in 1980 and 1981. We were part of a small group of sometimes-maligned supporters of the historic wilderness trails in southwestern BC. Both trips were challenging, rewarding, strenuous and certainly unforgettable. "Easy" is one description that wouldn't fit for the seven-day hike across this northern spur of the Cascade Mountains.

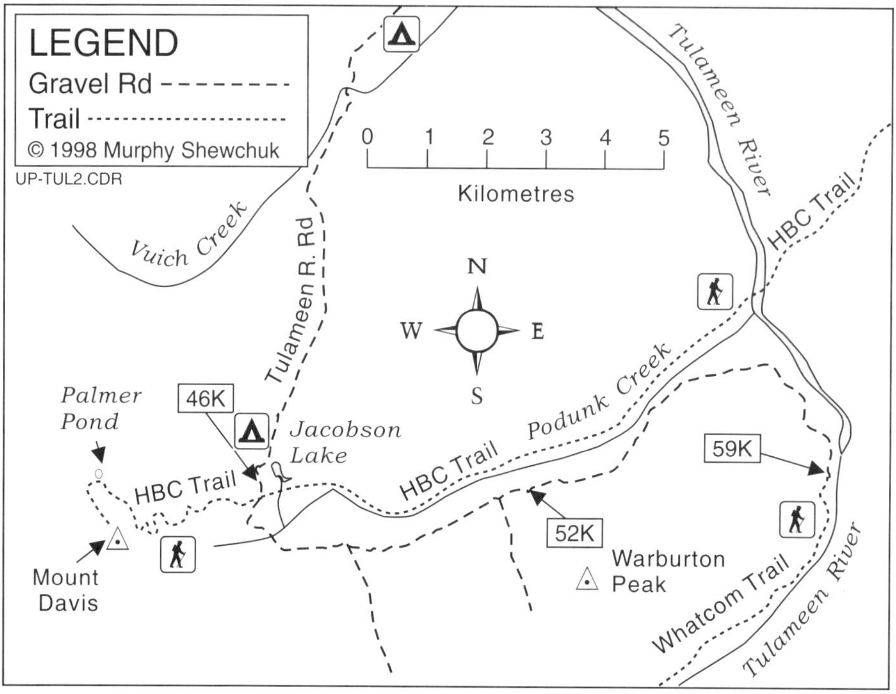

Map 7
Jacobson Lake and Upper Tulameen River area.

Now it should take you less than an hour to drive into the heart of the trail system — in dry weather, of course. Leave Coquihalla Highway 5 at the Coquihalla Lakes Interchange (Exit 228), then follow the signs to the Britton Creek Rest Area before continuing south for about 0.2 kilometres to the start of the Tulameen River Forest Road. The junction is at the site of an old Kettle Valley Railway (KVR) "Y" where the engines were turned around or parked after pulling the freights up the steep Coquihalla Canyon from Hope. By using the higher Boston Bar Creek valley, Coquihalla Highway 5 bypasses the toughest and most scenic section of the old railway route.

Use caution, particularly on weekdays, as Tulameen River Road is a well-used logging road. For the purposes of this backroad adventure, the junction at the "Y" near the north end of Coquihalla Lake is km 0. The road soon swings to the east of Coquihalla Lake into huckleberry country. Logging and forest fires, some of them caused by the fire-belching KVR steam locomotives, kept the forest from squeezing out the shrubbery. Newspapers of half a century ago report that the Natives picked hundreds of pounds of huckleberries near Coquihalla Pass.

Look closely at the brush along the roadway and you just might see the flicking ears of a mule deer.

The first major junction is near km 8.2, where the Britton Creek Forest Service Road continues east past the Murphy Lakes and over the mountains to Tulameen. (See *Coquihalla-Tulameen Road*, page 41.) The Tulameen River Road swings to the right, crosses Britton Creek and winds its dusty way southward through several logged areas.

Side roads in the area are confusing because the logging activity often makes them appear better than the main route. The kilometre numbers near the road should be watched carefully. An unexpected change in sequence could mean you have left the main route.

One of these side roads, near km 10.5, leads up and south to one of the access trails into Coquihalla Mountain. Keep left at all the junctions, and if you aren't stopped by washouts, you should reach the end of the road near a creek. Cross the creek, and you should intercept a trail to the alpine. Corrals on the right hand side of the main road near km 12.8 mark the foot of this same trail. A de-activated road just south of the Illal Creek bridge (km 19) also heads west into the same area.

A junction near km 24 marks the start of the road down the Tulameen Canyon to Tulameen. Keep to the right at km 25, continuing past the Champion Creek Road junction near km 26. The Tulameen River Road continues south, passing several logged areas.

Railroad Creek, at km 32, is one of many boulder-strewn waterways in the area. The creek, according to Barrie Sanford in *McCulloch's Wonder* (pp. 55-57), was part of one of three railway routes from Hope to Princeton put forward by Edgar Dewdney in 1901. This route would have followed the Tulameen River upstream from Princeton, through the village of Tulameen, and up through the canyon to Railroad Creek before crossing Railroad Pass and following Unknown (Carry) Creek down to the Coquihalla River. Although considerably shorter than the route adopted a decade later, the grades were just too steep for Railroad Pass to receive serious consideration.

Although the change is not obvious at first, the forest road leaves the Tulameen River halfway between Kelly Creek and Railroad Creek and begins following Vuich Creek into the heart of the Cascade Mountains. According to signs near km 38, hunting with a vehicle is prohibited beyond this point. The road crosses Vuich Creek and begins a steady climb through an old burn to Jacobson Lake at km 45.5.

The Tulameen River Road crosses the Brigade Trail near km 46. If you are interested in a good hike, take your lunch and a water bottle and explore the trail to Palmer Pond on Mount Davis. (See *Mount Davis Brigade Trail*, page 49.) After a swim in the cool, clear waters of Palmer Pond — elevation 1830 metres — you can retrace your steps to the Jacobson Lake Forest Service recreation site.

After crossing Podunk Creek near the HBC Brigade Trail junction, the road swings east and down the Podunk Creek valley, keeping well away from the creek. At the time of writing, the road ended in a large clearing near km 59.

A search in the trees at thc south end of the clearing near km 59 will reveal the remnants of a well-used trail. There are actually several old trails in this area. The oldest of these is Blackeyes Trail, a Native trading route to the Fraser Valley. The Whatcom Trail, a short-lived attempt by American merchants to cash in on the 1858 Gold Rush, also passes alongside the Tulameen River. Later the Dewdney Trail was cut through here to the lower Similkameen. Built in 1860-61, part of this route went from what is now known as Rhododendron Flats on Highway 3, across the Cascades to Whipsaw Creek and down to Princeton.

I hiked three kilometres upstream along the trails on the north side of the Tulameen River early in the summer of 1995. Although stranded in places, this section of the trail, marked variously as Whatcom and Dewdney, was obviously well-used. One section winds down to the Tulameen River and then through a marshy area before climbing away from the river. I tried a different route coming back, but ran into a blockage on the edge of a canyon. I had to retrace my route part way back to a steep descent to the marshy area. It's obvious that others have explored the route — I hope to do so in the near future.

•••

GPS References for major points of interest. See page 9 for details.

Wpt	Description	Latitude	Longitude
W01	Tulameen River FS Rd Start	N49° 38' 56.045"	W121° 00' 03.364"
W02	Lawless-Britton Cr. FS Rd	N49° 35' 20.176"	W120° 59' 11.589"
W03	Access To Coquihalla Mtn Trail	N49° 34' 34.851"	W120° 58' 16.897"
W04	Corrals at Coquihalla Mtn Trail	N49° 33' 45.316"	W120° 57' 06.757"
W05	Illal Creek Bridge	N49° 31' 53.365"	W120° 57' 44.743"
W06	24k New Jct. to Tulameen	N49° 29' 56.683"	W120° 56' 47.696"
W07	25k Old Jct. to Tulameen	N49° 29' 34.319"	W120° 56' 45.398"
W08	Tulameen-Champion FS Rd	N49° 29' 10.392"	W120° 57' 25.605"
W09	Tulameen-Railroad FS Rd	N49° 27' 04.227"	W121° 00' 00.409"
W10	Railroad Cr. Bridge	N49° 27' 01.697"	W121° 00' 11.707"
W11	Vuich Falls Rec. Site	N49° 26' 07.663"	W121° 00' 40.153"
W12	Sutter Cr. Rec. Site	N49° 26' 04.244"	W121° 00' 48.303"
W13	Treasure Mtn Rd	N49° 24' 52.810"	W121° 02' 06.748"
W14	Vuich Cr. Bridge	N49° 24' 29.365"	W121° 02' 10.630"
W15	Jacobson Lake Rec. Site	N49° 21' 11.012"	W121° 03' 29.499"

•••

10

Mount Davis HBC Brigade Trail

Statistics	For maps, see pages 45 and 50.
Distance:	4.5 kilometres, Jacobson Lake to Palmer Pond.
Hiking Time:	Two to three hours each way.
Elevation Gain:	300 to 350 metres.
Condition:	Some steep, muddy sections.
Season:	Best in late August.
Topo Maps:	Hope 92 H/6 (1:50,000).
Forest Maps:	Merritt Forest District Recreation Map.
Communities:	Merritt & Hope.

The Mount Davis section of the HBC Brigade Trail is arguably the most scenic and pristine part of what remains of this major commercial route. Over this trail, from 1849 to 1860, horse brigades carried the furs collected from as far north as Stuart Lake. They were bound for Qua-Qua-Alla, the present site of the town of Hope and the gateway to water transport to the Pacific Ocean.

Accompanying the Kamloops Brigade in July 1855 was Paul Fraser, Chief Trader from Fort Kamloops. According to several reports, Fraser was an autocratic tyrant with an uncontrollable temper. His ruthlessness was probably influenced by his association with Chief Trader Donald Manson, a man also noted for his ill-treatment of the men.

Earlier in 1855, Fraser had severely beaten one of the men at Fort Kamloops, a French Canadian named Farlardeau. The man died two days after the beating.

Two months later Chief Factor James Douglas received a letter from Donald Manson, dated July 7th, 1855, which contained in part "...I have at this moment (1 p.m.) arrived here from Qua-Qua-Alla, where I left Mr. McLean at 6 a.m. this morning with our Brigade. I am sorry to inform you of the death of an able and worthy friend - Chief Trader Paul Fraser. Poor gentleman, he only survived the blow about an hour. He never spoke. I shall see that he is interred as well as our means permit."

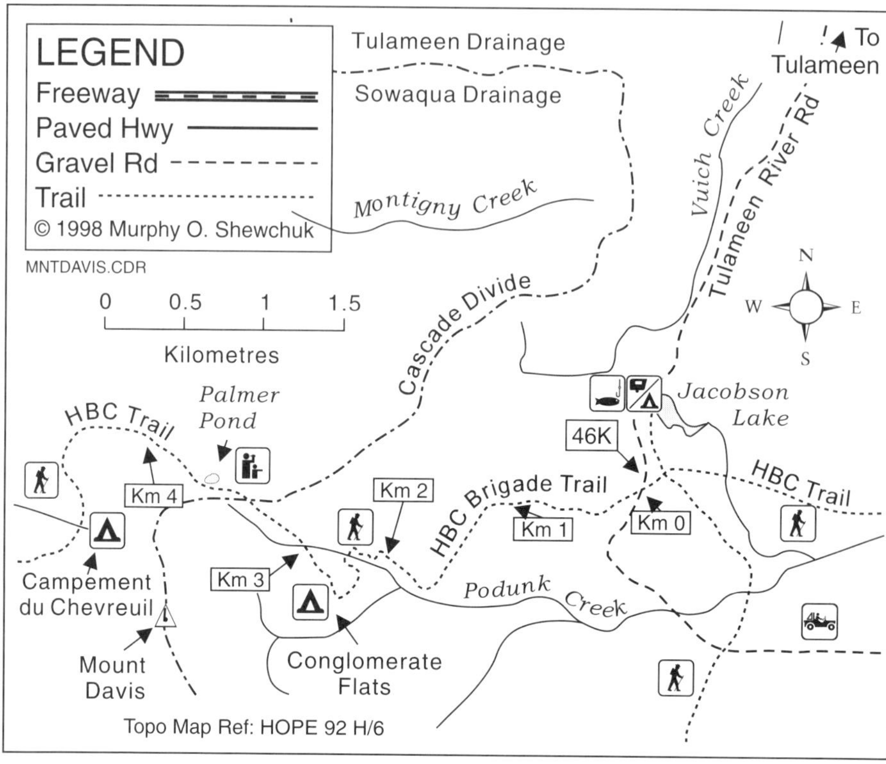

Map 8
The HBC Brigade Trail in the Mount Davis area.

Chief Factor James Douglas's report to the Hudson's Bay Company's secretary dated July 25th, 1855 adds more detail. "Letter arrived from Langley announces the death of Chief Trader Paul Fraser at the *Encampment de Chevruil.* A tree fell on him and he survived only about an hour."

There is some doubt that the death, which took place at their camping place on the western slopes of what is now called Mount Davis, was purely accidental. According to some reports, Paul Fraser was in his tent resting after a hard day on the trail. The men were continuing to prepare the camp for the evening, when the tree being cut for firewood or bed boughs fell on the tent. It narrowly missed Donald Manson who was also on that particular Brigade. From the standpoint of the men, neither man would have been mourned for long.

Podunk Davis, whose name is now attached to the mountain and a creek that drains the alpine meadows, was a major participant in a story that had a happier ending — at least where the Cascade Trails are involved.

According to D.M. Waterman, in an article in the 1976 *40th Report of the Okanagan Historical Society* (Nurse Warburton, pp. 73-74), Mary Warburton was a nurse who worked in Vancouver in the winter months and picked fruit in the Okanagan Valley during the summer. Nurse Warburton was determined to hike the Hope-Princeton (Dewdney) Trail and was not easily deterred by the saner heads around her.

In 1926, Mary Warburton set out on foot from Hope. After ten days her sister in Vancouver became alarmed when she received no word of her arrival in Princeton. The police were alerted and search parties were formed at Hope and Princeton. Over two weeks passed with no sign of the missing nurse, and all hopes faded.

Suddenly the word spread that she had been found. When she explained her circumstances to D.M. Waterman, she told of being led astray by well-used game trails and of finding an old trapper's cabin just when she thought all was lost. After resting, she had set out again in search of the trail when "I suddenly saw a white-haired figure approaching me through the trees... my first thought was that it was the Angel Gabriel..."

The "Angel Gabriel" was none other than Podunk Davis, an impressive looking trapper and guide, who had continued the search for Miss Warburton on his own after all hope had been abandoned. For his persistence and ultimate success, Podunk Davis was awarded the Royal Canadian Humane Association Medal.

A year or more later, Nurse Warburton succeeded in conquering the Hope-Princeton Trail. However, it is a fitting finale to Miss Warburton's story that she vanished with scarcely a trace on the Squamish-Indian River Trail in October, 1931.

The HBC Brigade Trail from Jacobson Lake to Palmer Pond, at the crest of Mount Davis, and down to *Campement du Chevreuil* should pose few of the hazards that Mary Warburton faced — although game trails are plentiful.

The trail begins 100 metres past the 46K marker on the Tulameen River Road, less than a kilometre south of Jacobson Lake. (See *Tulameen River Road*, page 45.) There is little room to park near the trail head, so vehicles are best left at a gravel pit near the recreation site at Jacobson Lake. The starting elevation is about 1500 metres (according to my pocket altimeter) and the trail climbs steadily westward from the road as it skirts a semi-open rocky knoll. A fire, believed to have taken place in the 1930s, has stripped much of the old-growth timber from this slope, leaving new growth interspersed with a carpet of huckleberry bushes and other shrubs.

Figure 11
Marmots frequent the slopes above Conglomerate Flats. (Photo MOH-043)

After about 1.5 kilometres, the trail gradually descends to a small creek and a meadow before beginning the serious climb up to Conglomerate Flats. Before reaching the Flats, the trail crosses the north fork of Podunk Creek (elevation 1650 metres) just below a rocky canyon. Then the trail swings south and up into a beautiful sub-alpine meadow dotted with large conglomerate (puddingstone) boulders. Marmots, pikas and deer — and maybe a few elk — make this meadow home. It is also an excellent location for a lunch break or wildflower photography break. In mid-August, 1990, we identified rare butterwort (Pinguicula vulgaris), an insectivorous violet-like flower plus fields of monkeyflower, Indian paintbrush and red mountain heather.

The trail skirts the meadow to the right (northeast), before cutting through a narrow stand of timber and into a sloped snowslide path. At the three kilometre point it emerges at the foot of a long narrow meadow just above the rocky canyon that carries Podunk Creek. The trail may be a little difficult to trace as it winds up the right (north) side of the meadow through a waist-deep field of hellebore, but the destination is Palmer Pond, just over the horizon at the top of the meadow. Quietly approach the crest of the meadow, and you may see deer on the slopes.

Palmer Pond is named after Lt H.S. Palmer, RE. He kept a detailed diary of his trip through the area in September 1859. The lake is beautifully clear and round and — at 1800 metres — understandably cool should you decide to take a swim. A game trail through the juniper thicket near the outlet of the lake leads to an excellent view of Montigny Creek, almost 600 metres straight down. To the north lies the red slopes of 2156-metre-high Tulameen Mountain.

Figure 12
A hiker at Palmer Pond on Mount Davis. (Photo BRD-059)

The trail climbs another 50 metres or so as it angles across the slope above the lake. It then continues through the alpine meadows, keeping to the right around the north side of the mountain. After leaving the meadows, it winds through the timber on a 200 metre descent along a hogs-back ridge to Chevreuil Creek and *Campement du Chevreuil.*

If you're not in a rush, you could stash your pack and follow the game trails south along the ridge to the top of Mount Davis. It could take you one-to-two hours to reach the peak and half that to return to the lake.

If you are continuing to *Campement du Chevreuil*, the trail through the alpine and along the edge of the cliffs above Palmer Pond is easy to lose. Generally keep to the right and watch for survey ribbons and red metal flags on the trees. Horses are occasionally used on the trail, so hoof prints and horse manure are sure signs that you are still on the trail.

It is easy to understand why *Campement du Chevreuil* was one of the favored overnight stops for the Brigades. The clear stream flows down off Mount Davis, even in mid-summer and there is plenty of shelter in the trees and plenty of grass in the meadows.

In mid-August, the open slopes to the north of the camp are an extended berry patch of giant proportions. I found time to enjoy the afternoon sun and nibble away on huckleberries, low bush blueberries and wild strawberries — all from one resting spot.

Beyond *Campement du Chevreuil,* the Brigade trail winds westward across the slope before making a switchback descent to Sowaqua Creek near the mouth of Matthew Creek.

A logging road has been built up the Sowaqua valley from Coquihalla Highway 5, but locked gates restrict access. If you are not planning to continue on the Brigade Trail on what could be at least a two-day hike over Fool's Pass and Manson Ridge to Peers Creek, it may be best to retrace your steps to Jacobson Lake.

•••

GPS References for major points of interest. See page 9 for details.

Wpt	Description	Latitude	Longitude
W01	Jacobson Lake Rec. Site	N49° 21' 11.012"	W121° 03' 29.499"
W02	Start of Trail near 46K	N49° 20' 56.220"	W121° 03' 31.025"
W03	Conglomerate Flats	N49° 20' 37.661"	W121° 05' 14.266"
W04	Palmer Pond	N49° 21' 01.666"	W121° 05' 52.697"

•••

Figure 13
Andrea and Andrew Howes on July Mountain. See next page for details.
(Photo NVG-1273)

NEW

11

July Mountain Trails

Statistics	**For map, see page 58.**
Distance:	9 km, Juliet Interchange to parking area. Four to five km, parking area to July Mountain alpine.
Travel Time:	One half hour drive - 2 to 3 hour hike each way.
Condition:	Gravel road with rough sections.
Elevation Gain:	Approximately 1125 metres.
Season:	July through September.
Topo Maps:	Spuzzum 92 H/11 (1:50,000). Yale 92 H/NW (1:100,000).
Forest Maps:	Merritt Forest District Recreation Map.
Communities:	Merritt & Hope.

July Mountain is easily missed among the jagged peaks of the Cascade Mountains as you drive north on Coquihalla Highway 5. The barren peaks of Zopkios Ridge are much more impressive in the morning light. July Mountain presents a different picture, however, as you approach the Juliet Creek interchange when you are heading south. From this angle, the morning light illuminates an alpine basin hanging from the jagged semi-circular ridge.

An even better view of the mountain is from the crest of Thynne Mountain, directly east of the Juliet Interchange and accessible via Brookmere. The best view, of course, is from the mountain itself.

Juliet Creek Route

If washouts or road closures haven't blocked the way, getting to the top is relatively easy. Take the Juliet exit (Exit 239) and drive along the service road until you reach a T-junction near Juliet Creek. Then follow Juliet Creek FS Road upstream until you reach a bridge across the creek near km 4. (If you are interested, there is a campsite just before the bridge.) After crossing the creek, swing left (southeast) and follow the gravel road around the east flank of the mountain.

The road is steep in places and can be slippery when wet, but should pose no problems for most vehicles. Keep left at a junction near km 7.7 and again near km 8.2. By this time you should be heading west into the mouth of a long upland valley. There could be logging in the area and the safest place to park may be a clearing near the 9K marker.

Although an old logging road continues approximately two kilometres farther west into the July Mountain basin, it had been severely water-barred and would pose a hazard to most vehicles. Since you are undoubtedly out here for the quiet, beauty and exercise, find a safe place to park and lace up your boots.

The road/trail into the basin and up to the alpine keeps generally to the north of the creek, so keep right and angle up the hill, following the old road and skid trails two kilometres to the west end of the cut. Watch for survey ribbons and a recently cut trail through the old growth timber and underbrush. Although the last kilometre may convince you that you aren't in as good shape as you thought, it is only about two kilometres from the end of the logging to the alpine basin.

At the time of writing, there was some bushwhacking required to get through a short section of marshy timber and boulders. This section may have been completed by the time you get there. If it hasn't, keep to the base of the avalanche area and follow the west side of the creek.

Beyond the last of the big timber, there wasn't a clearly marked trail. However, we found that the easiest route to the alpine tarn and the main peak was to follow the dry creek beds up, keeping to the open slopes as we gradually swung south and into the mouth of the upland basin.

While the alpine tarn is surrounded on three sides by cliffs, there is a lower saddle to the west and a bit of side-hill gouging can take you up to the ridge and the last ascent to the 2125 metre peak.

July Creek Route

At the time of writing, another route was being explored via the Mine Creek interchange and the July Creek FS Road. This route presents an easier grade to a sub-alpine meadow and a shorter climb to the alpine ridges. However, it was little more than a flagged route through the underbrush.

To reach the "trail" head, take Exit 231 (Mine Creek Road). Note that Exit 231 is not accessible from the south — if you are travelling from the south, you will have to continue another seven kilometres north to the Juliet exit and then return south to the Mine Creek exit. With the Mine Creek interchange as km 0, follow the paved road north (up the hill), on the west side of the highway. The pavement ends at about 0.8 km, and July Creek Road swings to the right at km 1.2. Follow it uphill, past an abandoned mine, until you come to a three-way junction near the 2K marker (about 3.2 kilometres from the highway). Take the middle fork and continue up the base of the July Creek valley.

At the time of writing, the logging road ended about 7.5 km from the highway and the hiking route continued southwest for about one kilometre before reaching a large sub-alpine meadow. A skree slope to the south of the meadow provides access to the ridge. Game trails to the west also may prove to be a viable route to the ridge. Once on the ridge, it is a relatively easy hike north to the main peak.

•••

GPS References for major points of interest. See page 9 for details.

Wpt	Description	Latitude	Longitude
	Juliet Creek Route		
W01	Start Juliet Cr. FS Rd	N49° 44' 18.285"	W121° 00' 43.777"
W02	Juliet Cr. Bridge At 4K	N49° 44' 07.205"	W121° 03' 31.463"
W03	Jct. Keep Left	N49° 43' 19.818"	W121° 02' 04.540"
W04	Jct. Keep Left	N49° 43' 07.873"	W121° 02' 27.799"
W05	9K Marker	N49° 43' 03.889"	W121° 02' 41.145"
W06	Jct. Hike Upper Rd	N49° 42' 59.810"	W121° 03' 00.438"
W07	End of logging	N49° 42' 58.581"	W121° 04' 19.850"
W08	Marsh and Rockslide	N49° 42' 56.011"	W121° 04' 44.425"
W09	Alpine Tarn	N49° 42' 20.332"	W121° 05' 20.187"
W10	Alpine Ridge	N49° 42' 15.082"	W121° 05' 35.673"
	July Creek Route		
W11	Mine Creek Exit 231	N49° 40' 51.356"	W121° 00' 36.699"
W12	Start July Creek FS Rd	N49° 41' 17.066"	W121° 01' 01.007"
W13	Abandoned Mine	N49° 41' 21.349"	W121° 01' 10.965"
W14	Three-way Junction	N49° 42' 07.416"	W121° 01' 40.349"
W15	Huckleberry Heaven	N49° 42' 09.387"	W121° 01' 55.299"
W16	End of July Cr. Rd	N49° 41' 15.572"	W121° 04' 35.707"
W17	Subalpine Meadow	N49° 41' 09.334"	W121° 05' 19.478"
W18	Skree Slope	N49° 41' 02.653"	W121° 05' 20.838"

•••

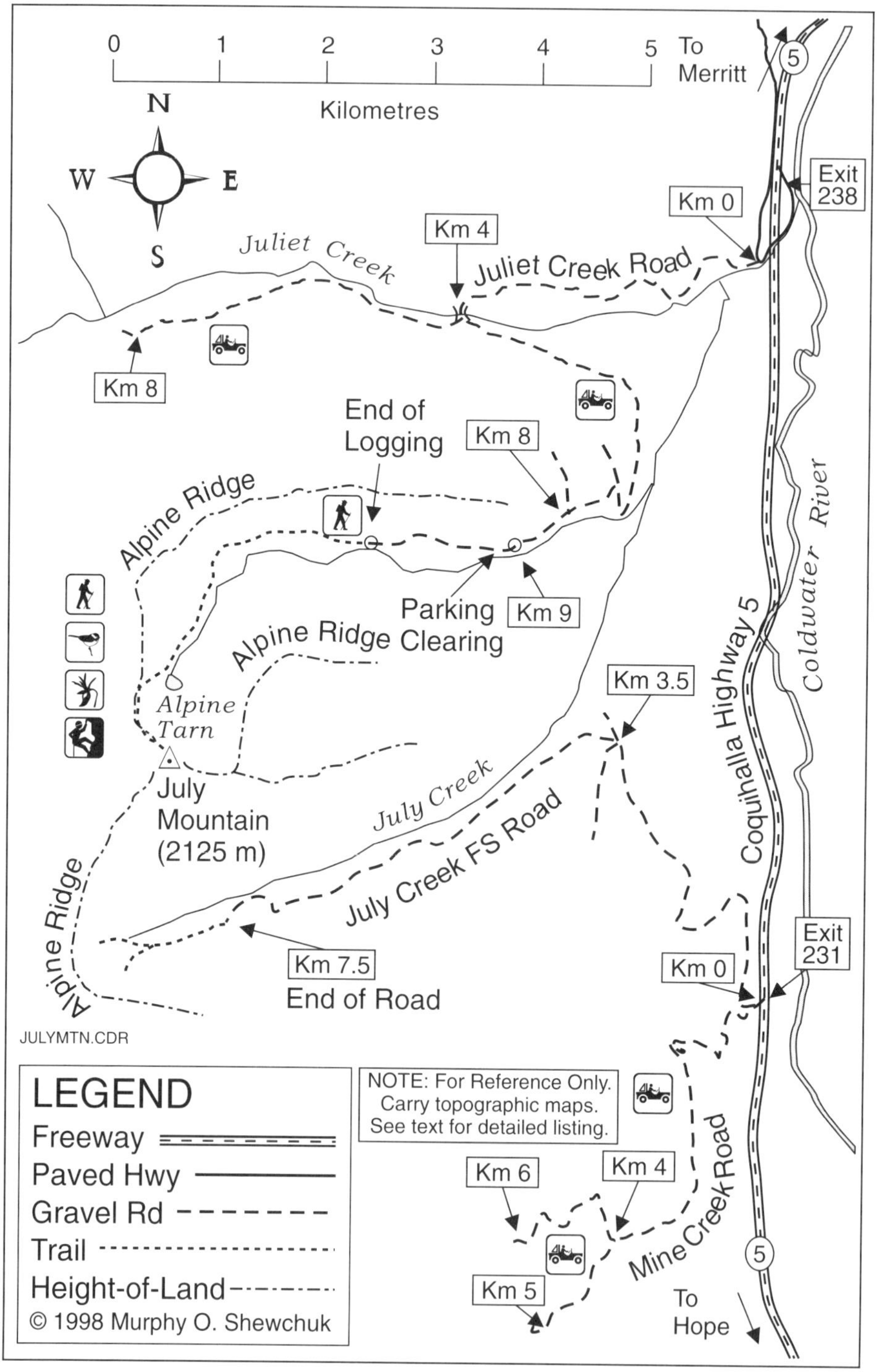

Map 9
July Mountain area.

12

Brookmere

Statistics	For map, see page 60.
Distance:	10 km, Coquihalla Highway 5 to Brookmere. 25 km, Coquihalla Highway 5 to Coalmont Road.
Travel Time:	15 min. to Brookmere. One hour to Coalmont Rd.
Condition:	Gravel with some rough sections.
Season:	Summer only, Brookmere to Coalmont Rd.
Topo Maps:	Aspen Grove 92 H/15 (1:50,000).
Forest Maps:	Merritt Forest District Recreation Map.
Communities:	Merritt & Hope.

Louis H. Brooks and Phillip P. Brooks were the first settlers in the community now known as Brookmere, registering their lots in 1909. In the spring of 1911, it appeared that both the Kettle Valley Railway (KVR) and the Vancouver, Victoria and Eastern Railway and Navigation Company (VV&E) planned to run rail lines from Princeton to the coast via the pass where the Brooks had their farms.

Plans were immediately made to survey a large new town on Louis Brooks' land. The agreement called for the railway company and Brooks to each get half of the 360 surveyed lots. Brooksville, it was said, would have a very advantageous position at the summit of Pass Creek (now Brook Creek) and Spearing Creek, the west fork of Otter Creek.

Brooksville became Brookmere, some say because of the swamp on which the town was built. It did become a divisional point for the KVR, but the major community never materialized. Today Brookmere has the distinction of having the last remaining KVR water tank still standing — although it had to be moved off its original foundation to save it from the wrecker.

Exit 256 (Coldwater Road) is the departure point for both Brookmere and the Coldwater Road. With the exit as your km 0 reference, head south on the Coldwater Road and you will cross two new bridges over the Coldwater River at km 1.5 and km 2.7. The 36K sign near km 3 refers to the distance to Merritt where most of the logs from this area are hauled.

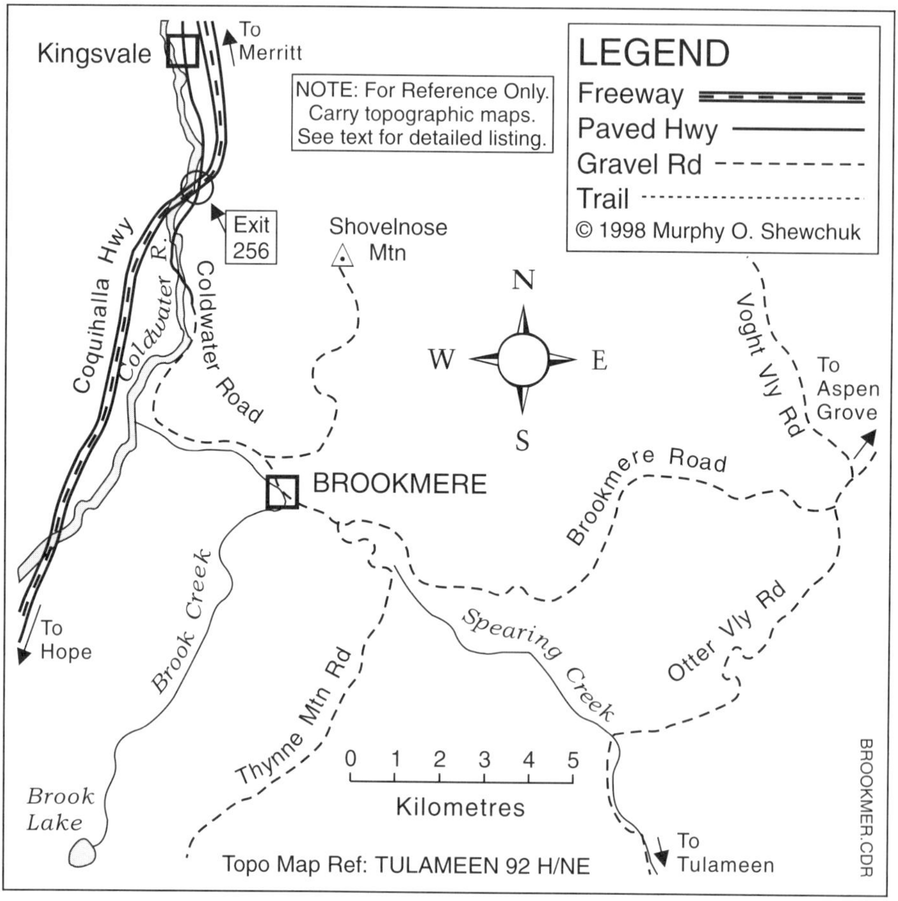

Map 10
Brookmere and area — from Exit 256 to Coalmont Road.

The south extension of Coldwater Road climbs steadily away from the Coldwater River after crossing the former KVR right-of-way near km 5. Moose Hill, one of the first steep climbs, has a much-deserved reputation for being treacherous in winter.

A side road at km 8.6 (near the 42K marker) heads up to the top of Shovelnose Mountain. There are a maze of forest roads on the mountainside, but access improvement for cellular telephone service may make the right road a little easier to find. It is about 9.3 km to the 1830 metre peak, and the view from the top can be worth the effort.

Figure 14
The Brookmere KVR water tank prior to the removal of the tracks.

The Brookmere water tank and caboose at km 10.2 mark the centre of the community. If you look carefully, you may see the foundation of a railway roundhouse and a large fuel storage tank, more reminders of the days when the railroad was the king of transportation.

Beyond the community, at km 11.5, a former railway crossing marks the start of a Forest Service road to the top of Thynne Mountain. (See *Thynne Mountain Road*, page 63.) This is also the start of the Thynne Mountain snowmobile trail network. Various work crews have built a loading ramp and cleared a parking area near the old railway crossing.

Brookmere Road crosses the right-of-way to the left and winds 15 km across the plateau, joining the Coalmont Road about 18 km south of Aspen Grove. See *Otter Valley Road* page 113.

Coldwater Road, from Merritt to the junction south of Brookmere is kept open year around. However, Thynne Mountain Road and Brookmere Road are normally only passable in summer.

•••

GPS References for major points of interest. See page 9 for details.

Wpt	Description	Latitude	Longitude
W01	Exit 256 - Highway 5	N49° 53' 01.770"	W120° 54' 05.976"
W02	KVR crossing	N49° 51' 13.176"	W120° 54' 27.798"
W03	Shovelnose Mtn Road	N49° 49' 38.394"	W120° 53' 11.922"
W04	Brookmere Watertower	N49° 49' 03.690"	W120° 52' 21.288"
W05	Jct. Coldwater & Brookmere Rds	N49° 48' 46.290"	W120° 51' 23.892"
W06	Old decaying buildings	N49° 48' 30.534"	W120° 46' 24.828"
W07	Jct. Coley Creek Road	N49° 49' 17.460"	W120° 46' 04.086"
W08	Jct. Coalmont Road	N49° 48' 58.554"	W120° 42' 18.384"

•••

13

Thynne Mountain Road

Statistics	For maps, see pages 60, 65 and 70.
Distance:	23 km, Brookmere Road to Thynne Peak.
Travel Time:	Approximately one half hour.
Condition:	Gravel with rough sections.
Season:	July through October.
Topo Maps:	Aspen Grove 92H/15 (1:50,000). Tulameen 92H/10 (1:50,000).
Forest Maps:	Merritt Forest District Recreation Map.
Communities:	Merritt & Hope.

One of Thynne Mountain's attractive features is that it is an accessible alpine ridge where you can see alpine plants typical of the North Cascade Mountains. Countless western anemone blossom in the wake of the retreating snow, then thrust their "towhead baby" seed pods up to catch the wind. Avalanche lilies present their yellow blossoms to the sun, then shyly fade away for another summer. Colourful blue, pink and white phlox hug the ground to avoid the incessant breeze. Many other varieties of wildflowers, some of them stunted by the short summers, send forth their blossoms, then retreat into hiding.

Deer feed on the alpine growth during the short summer months while cattle graze on the lower slopes. Sheep also once roamed the alpine meadows. Bears enjoy the lush growth at the lower elevations and in the autumn can be sometimes seen reluctantly sharing the huckleberries with marmots and chipmunks.

The view from the top of Thynne Mountain encompasses the grassland plateau to the north and east, and the Coast Range and Cascade Mountains to the west and south. If the weather is clear, the snow-capped top of Mount Baker can be seen to the south through a "V" created by the Coquihalla Canyon.

The rough gravel road to the top of the 2020-metre-high ancient volcanic peak services two microwave repeater sites and a B.C. Forest

Service Fire Lookout. Two excellent fishing lakes are also nestled on the north side of the mountain only a short distance from the road (see *Andys Lake Loop*, page 67).

Figure 15
Thynne Mountain in winter.

Caution is necessary when travelling this road on weekdays as logging may be taking place on the lower elevations of the mountain. The trucks usually start hauling very early in the morning, but usually make their final trip before 5:00 pm.

The kilometre signs used by the truck drivers can also help you through the maze of roads. The starting point, km 0, is at the old railway crossing approximately 1.5 kilometres southeast of Brookmere. Thynne Mountain Road does not cross the former right-of-way but keeps to the south, gradually climbing through the timber. Keep straight ahead at the Brook Creek Road junction near km 2.0 and to the right at a junction at a new log landing at km 4.8. Swing to the right at the Lower McPhail Road junction at km 6.8.

From this point, small red flags and yellow plastic tubes fastened high up trees and posts mark the route to the AT&T (Unitel) communications site at the top. Maintenance technicians find these flags particu-

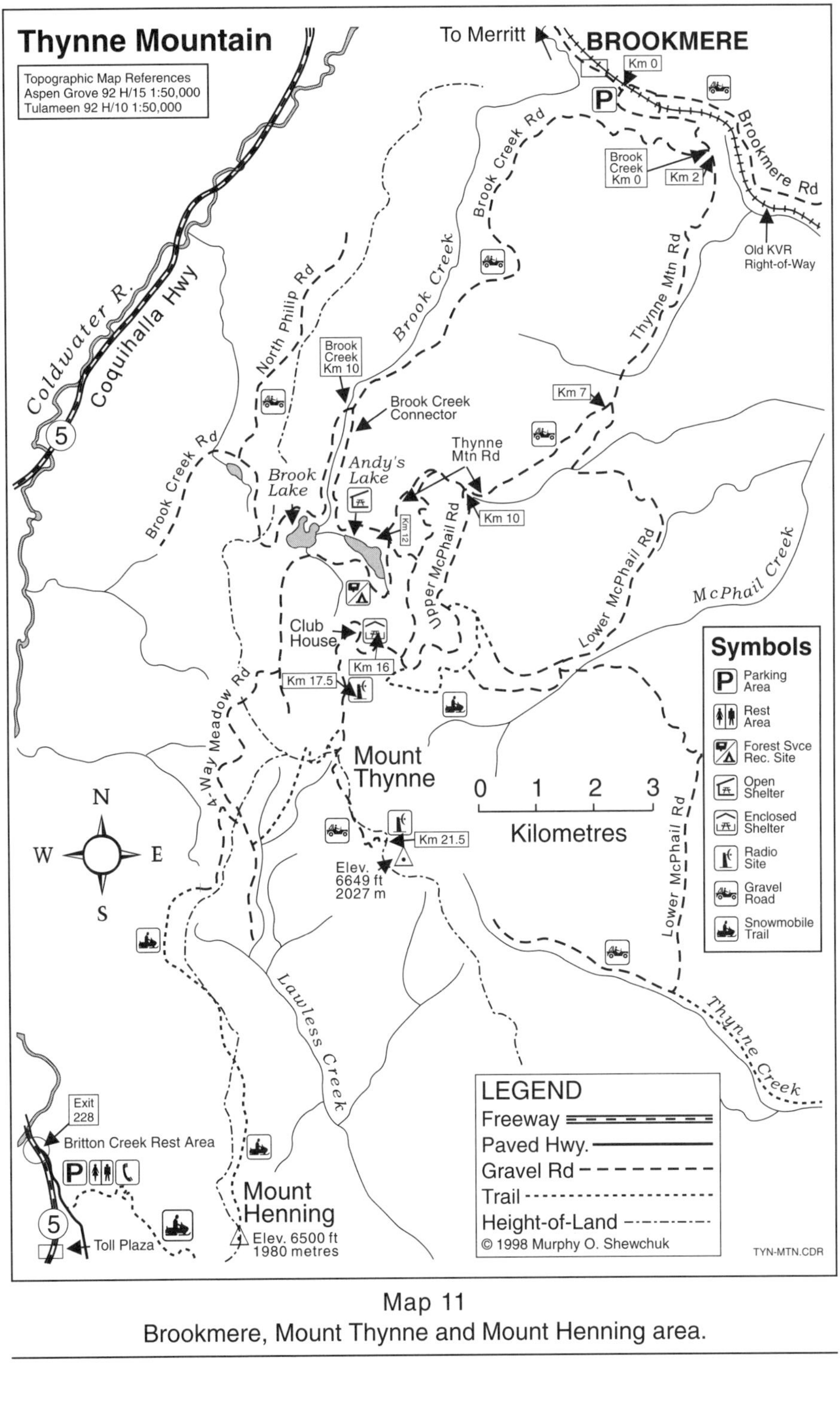

Map 11
Brookmere, Mount Thynne and Mount Henning area.

larly useful in winter when storms and deep snow obscures many of the usual landmarks.

You have two options at an important junction at km 10. If you take the Upper McPhail Road to the left, keeping right at the next two junctions, you will rejoin the Thynne Mountain Road near km 15 and avoid several loose, steep grades.

If you keep right at Upper McPhail Road, at the junction at the 11K sign take the lesser used road to the left. The road to the right heads into Andys Lake while the left road proceeds across a small flats and begins an undulating climb to the top.

The Merritt Snowmobile Club has converted an old CN-CP generator building to a clubhouse at km 16 and, half a kilometre later, the steep, rough climb to the crest of the ridge begins. The road levels off at a communications site at km 17.5. It then winds among the meadows on the top of the ridge for approximately five kilometres to the CN-CP (Unitel/AT&T) communications site.

There are several small, shallow lakes and ponds nestled in the folds of the alpine ridge. Wilderness camping options abound and the wildflower meadows can be spectacular in July and August.

•••

GPS References for major points of interest. See page 9 for details.

Wpt	Description	Latitude	Longitude
	Thynne Mountain Road		
W01	Start Thynne Mtn Rd	N49° 48' 51.948"	W120° 51' 18.348"
W02	Jct. Brook Cr. Rd	N49° 48' 17.574"	W120° 50' 37.020"
W03	Jct. Lower McPhail Rd	N49° 46' 04.494"	W120° 51' 59.634"
W04	Jct. Upper McPhail Rd (10K)	N49° 45' 18.378"	W120° 54' 06.900"
W05	Upper McPhail Cutoff	N49° 43' 40.488"	W120° 54' 54.564"
W06	Jct. Old Thynne Mtn Rd	N49° 43' 47.688"	W120° 55' 10.590"
W07	Snowmobile Club Building	N49° 43' 49.044"	W120° 55' 39.060"
W08	Communications Site	N49° 43' 36.120"	W120° 55' 53.946"
W09	Alpine Pond	N49° 43' 03.738"	W120° 55' 55.800"
W10	Alpine Pond	N49° 42' 12.984"	W120° 55' 43.092"
W11	Communications Site	N49° 42' 08.082"	W120° 55' 19.884"

•••

14

Andys Lake Loop

Statistics	For maps, see pages 65 and 70.
Distance:	Approximately 18 km round trip.
Travel Time:	one half to one hour.
Condition:	Summer road, used by snowmobilers in winter.
Season:	Summer and Winter.
Topo Maps:	Aspen Grove 92 H/15 (1:50,000).
	Tulameen 92 H/10 (1:50,000).
Forest Maps:	Merritt Forest District Recreation Map.
Communities:	Merritt & Hope.

Thynne Mountain, particularly the Andys Lake area, is of recreational interest both winter and summer. The only time of the year that the wildlife have the area much to themselves is in spring when mud and mosquitoes keep the less-determined fishermen at home. Andys Lake is also a destination on Merritt Snowmobile Club's Thynne Mountain circuit.

Because the snowmobile route differs somewhat from the roads when winter logging is active, the following directions will apply primarily to summer traffic. Snowmobilers should check the maps posted at the starting point near Brookmere or obtain a copy of the snowmobile route map from the Forest Service in Merritt.

For the first 10 kilometres, the directions are the same as Thynne Mountain Road. The starting point, km 0, is at the former KVR crossing approximately 1.5 km southeast of Brookmere. Thynne Mountain Road does not cross the former right-of-way but keeps to the south, climbing through the timber. Keep straight ahead at the Brook Creek Road junction near km 2.0 and to the right at a junction at a log landing at km 4.8. Swing to the right at the Lower McPhail Road junction at km 6.8.

Keep right at the Upper McPhail Road/Thynne Mountain Road junction at the 10K marker and again near 11K. The road to the right heads into Andys Lake while the old Thynne road proceeds across a small flats and begins an undulating climb to the top.

Figure 16
Nola, Matthew and Bridget explore an alpine pond on Thynne Mountain. (Photo THY-173)

Keep left at a little used road at km 12.1 and watch for a junction at 12.4 where the road to the right dips quickly down a short hill before disappearing around the corner. Remember this junction as it provides an alternate route out.

Andys Lake North Recreation Site, at km 13, serves as a base for fishing and hiking in the area. Beyond the recreation site, the road gradually descends to a creek crossing and another small recreation site nearer the lake. A rough road named "Crosser" near km 13 links Andys Lake with Thynne Mountain Road. There are trails through the trees and a shelter near the northeast shore of Andys Lake.

From the junction at km 12.4, the loop continues north and then northeast as Brook Creek Road. Keep an eye out for a trail into Andys Lake less than half a kilometre farther along. A major road comes in from the left about 3.4 km from the junction, and then its all downhill to Thynne Mountain Road, about three kilometres from Brookmere.

Note: See *Thynne Mountain Road*, page 63, for GPS references.

•••

15

Thynne-Henning Trail

Statistics	**For maps, see pages 65 and 70.**
Distance:	7 km, trailhead at km 24.5 to Mount Henning summit.
Travel Time:	Five to six hours round-trip hiking time.
Condition:	Old sheep trail, used by snowmobilers in winter.
Season:	Summer and Winter.
Topo Maps:	Tulameen 92 H/10 (1:50,000).
Forest Maps:	Merritt Forest District Recreation Map.
Communities:	Merritt & Hope.

There is an old sheep trail that starts on the west face of Thynne Mountain, three km west of the south summit, and winds south through semi-open timber and alpine meadows to Mount Henning. It is undeveloped and, although used by snowmobilers in winter, may be difficult to find it in summer. When we last hiked it, it was flagged with survey ribbons and showed signs of recent ATV traffic.

It begins at the end of the road near km 24.5 and winds southwest for about 1.3 km to what we called Eagle Feather Lake. After passing on the east side of the lake, it descends through a meadow for a short distance then heads east for about 0.5 km. It continues southeast for about one kilometre to a large marsh and then southward generally along the east side of the ridge to the summit of Mount Henning.

GPS References for major points of interest. See page 9 for details.

Wpt	Description	Latitude	Longitude
W01	End of road / trailhead at km 24.5	N49° 41' 37.836"	W120° 57' 52.776"
W02	Eagle Feather Lake	N49° 41' 06.090"	W120° 58' 33.810"
W03	Trail turns south	N49° 40' 58.718"	W120° 58' 12.027"
W04	Large marsh	N49° 40' 31.578"	W120° 57' 50.784"
W05	Open meadows	N49° 40' 11.940"	W120° 57' 45.048"
W06	Jct. keep left	N49° 40' 07.014"	W120° 57' 32.904"
W07	The Wall north edge	N49° 39' 14.466"	W120° 57' 28.728"
W08	Henning ponds	N49° 38' 50.772"	W120° 57' 39.948"
W09	Henning summit	N49° 38' 39.858"	W120° 57' 28.536"

•••

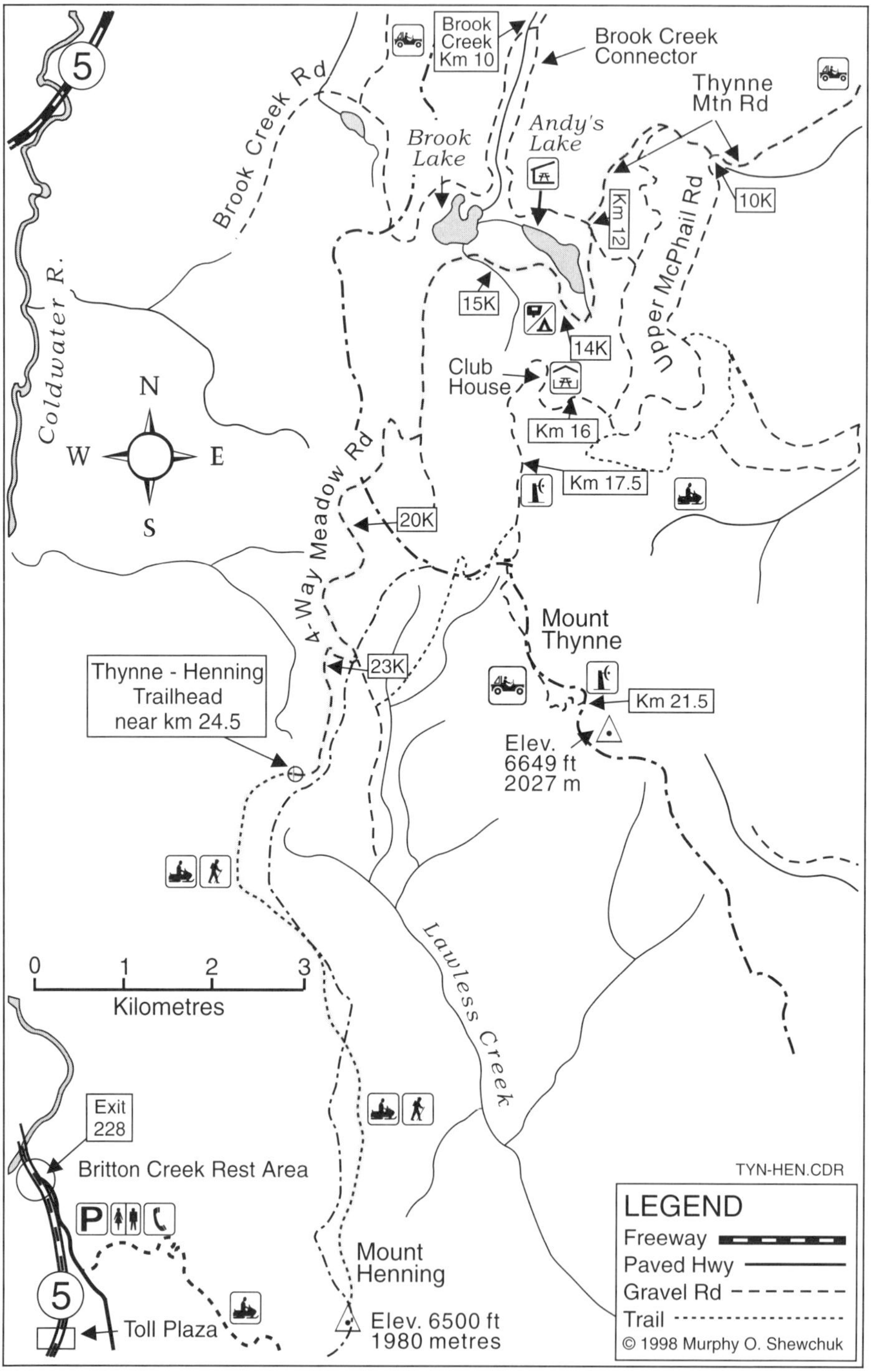

Map 12
Mount Thynne - Mount Henning Trail.

16

Coldwater Road

Statistics	**For maps, see pages 4 & 59.**
Distance:	31 km, Exit 256 to Highway 5A/97C at Merritt.
Travel Time:	One half hour.
Condition:	Paved.
Season:	Year around.
Topo Maps:	Tulameen 92 H/NE (1:100,000). Merritt 92 I/SE (1:100,000).
Forest Maps:	Merritt Forest District Recreation Map.
Communities:	Merritt.

The original highway through the Coldwater Valley followed the benches above the river, well below Coquihalla Highway 5. When survey work started on the Coquihalla Highway, local ranchers and Indian Bands objected to the potential destruction of prime rangeland by a multi-lane divided highway. Their voices were heeded and the new highway was routed along the mountainside, above the agricultural land. With the Coldwater Road Interchange (Exit 256) as km 0, the first diversion on this paved backroad is Kingsvale at km 2.7.

Kingsvale does not have a connection to royalty, but instead was named after Del King. King, son-in-law of Nicola Valley pioneer Jesus Garcia, settled on land where a railway station and post office were later located. High hopes for a long-term settlement disappeared shortly after World War I when Major Goldman, then owner of the Nicola Ranch near Merritt, bought up the main ranching properties in the nearby Voght and Kane valleys. The former KVR/CPR underpass at Kingsvale marks the junction of a backroad west to Gillis and Murray lakes. (See *Murray Lake Road*, page 73.)

Kane Valley Road, at km 4, provides access to an excellent cross-country skiing area, a series of fine upland lakes, and, if time is no concern, a shortcut to Highway 5A/97C near Corbett Lake.

(See *Kane Valley Road*, page 83, for details.)

Patchett Road, at km 12.6, provides access to the upper reaches of Spius Creek and a convoluted backroad to Boston Bar, in the Fraser Canyon. The shortcut to Boston Bar is difficult to find, even more difficult to describe, but see page 77 for my attempt at it.

The Coldwater Indian Reserve, at km 19, is home to well-known local Native artist, Phillip "Opie" Oppenheim. Opie has operated a studio and gallery here since 1987. His excellent work, in silver, jade and silk-screen prints, is much in demand.

Figure 17
Stocking steelhead in the Coldwater River near Kingsvale. (Photo CLD-051)

Veale Road, near km 26, is the starting point of a gravel road that leads up to the top of 1693-metre-high Iron Mountain and also to a fine fishing lake on the south slopes of the mountain. (See *Gwen Lake Road* on page 93.)

Thirty-one kilometres after leaving Exit 256, the Coldwater Road joins Highway 5A/97C, about two kilometres southeast of downtown Merritt.

•••

17

Murray Lake Road

Statistics	For map, see page 75.
Distance:	33 km, Kingsvale to Juliet via Murray Lake.
Travel Time:	One to two hours in dry weather.
Condition:	Gravel road with rough sections.
Season:	Mid-May through October.
Topo Maps:	Tulameen 92 H/NE (1:100,000). Yale 92 H/NW (1:100,000).
Forest Maps:	Merritt Forest District Recreation Map.
Communities:	Merritt & Hope.

Gillis Lake, at the north end of this backroad, has an excellent Forest Service campsite and good trout fishing opportunities year around. The marshes at the southwest end of this boomerang-shaped lake can be quite productive early in the season. Although there are a few places to fish from shore, a canoe or cartop boat would be a useful accessory.

Murray Lake is a narrow 2.5-km-long lake lying in a north-south valley just west of the Coldwater River. It is a quiet, pretty lake suitable for canoeing or for fishing with a small boat and light tackle. There are small Forest Service recreation sites at both ends and a number of summer cottages on the east shore.

The southern access is from the Juliet exit on Coquihalla Highway 5, once a water stop on the Kettle Valley Railway. The road from Juliet is rough and climbs steeply for about five kilometres before reaching the south end of Murray Lake.

The northern access from Kingsvale, another old KVR water stop, is in better condition because of logging traffic. To reach Kingsvale, take Exit 256 on Coquihalla Highway 5, south of Merritt, and drive north for 2.7 kilometres to where the Coldwater Road passes under the former KVR right-of-way (30 kilometres south of Merritt on the Coldwater Road).

Figure 18:
Gillis Lake, five kilometres west of Kingsvale. (Photo CLD-072)

Murray Lake Road begins at the underpass so proceed straight west at the Kingsvale junction. Cross the Coldwater River at km 0.2 and swing south (left) on the road that begins climbing almost immediately. The road climbs steadily, twisting up the mountainside to Gillis Lake at km 5.0. There are a few hidden spots to park a vehicle or pitch a tent on the west shore of the lake, but the main recreation site is below the road on the north shore of Gillis Lake. The unmarked entrance to the Gillis Lake Forest Service Rec site is near the west end of the lake at km 6.4.

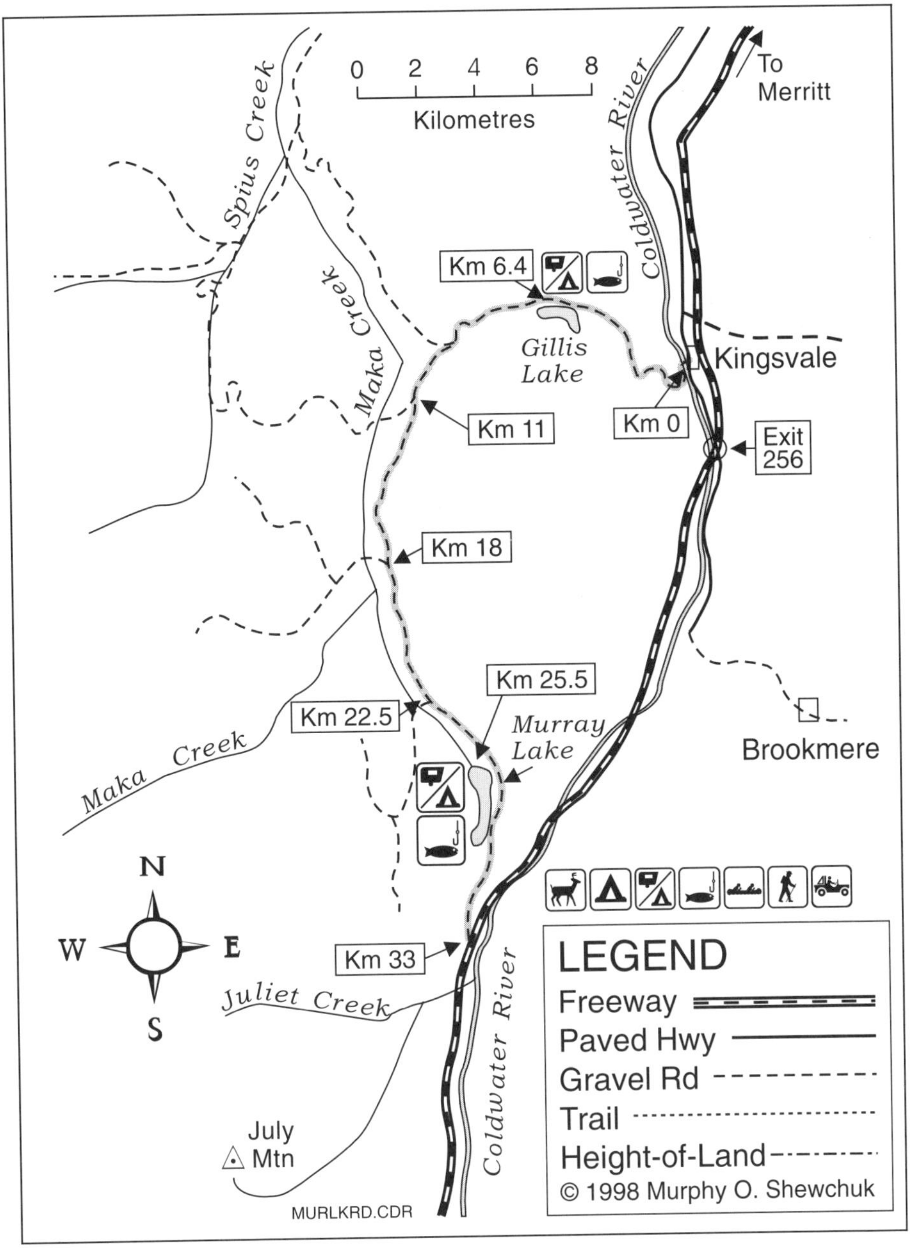

Map 13:
Murray Lake Road

Gillis Lake is named after Dr. Austin Gillis, who ran sheep in the area in the early part of the century. He and his brother, Dr. John Joseph Gillis, were active medical practitioners in the Nicola Valley for four decades beginning in 1911.

The Gillis Lake recreation site is not particularly large and, because of the sheltered lakeshore location, it fills up quickly on the weekends. However, on a fall weekday, you may have the opportunity to have the site and the trout-stocked lake all to yourself.

The Murray Lake Road then parallels a 500 kV power line to km 10 before swinging south and up Maka Creek. A junction at km 18 marks the Lower Maka Forest Service Road. Keep left to Murray Lake. In this area, the road passes through second growth forest and large stands of aspen. Deer and bear are a frequent sight, especially in the evening and early morning. Keep left again at the Maka-Murray Forest Service Road junction near km 22.0. The road to the right heads into an active logging area and several small lakes.

At a junction near the 25K signpost, the road swings right to Murray Lake while the road to the left continues to another logging operation. The Murray Lake North Forest Service recreation site near km 25.5 is a large, open site with a day-use area and a place to launch a cartop boat. The islands in the lake provide shelter and privacy from the limited traffic that passes this way.

The road past the cottages on the east shore of the lake appears to be designed to discourage traffic. It is rough, dusty and narrow and best approached with caution. The recreation site at the south end of Murray Lake is small with a little more shelter.

South of Murray Lake, the narrow, rough gravel road edges across the mountainside for three kilometres before making a two-kilometre switch-back descent to Coquihalla Highway 5 at Juliet Creek.

•••

GPS References for major points of interest. See page 9 for details.

Wpt	Description	Latitude	Longitude
W01	Kingsvale KVR Underpass	N49° 54' 21.138"	W120° 54' 46.008"
W02	Gillis Lake East End	N49° 55' 21.720"	W120° 57' 08.262"
W03	Gillis Lake West End	N49° 55' 26.238"	W120° 58' 01.158"
W04	Jct. to Spius Creek	N49° 54' 44.394"	W121° 01' 10.452"
W05	Jct. to Boston Bar	N49° 54' 39.948"	W121° 01' 15.744"
W06	Jct. Lower Maka FS Road	N49° 51' 24.666"	W121° 02' 41.892"
W07	Jct. North Michael Lake FS Road	N49° 49' 45.636"	W121° 02' 05.412"
W08	Maka - Murray FS Road	N49° 49' 17.496"	W121° 01' 37.914"
W09	Michael Lake	N49° 48' 28.932"	W121° 03' 23.298"
W10	Debbie Lake	N49° 47' 20.760"	W121° 02' 19.296"
W11	North Murray Lake Rec Site	N49° 47' 58.864"	W121° 00' 14.143"
W12	South Murray Lake Rec Site	N49° 46' 58.668"	W121° 00' 25.104"
W13	Coquihalla Highway 5	N49° 45' 12.642"	W121° 00' 43.686"
W14	Juliet Creek FS Road	N49° 44' 21.504"	W121° 00' 52.608"

•••

NEW

18

Kingsvale to Boston Bar

Statistics	For map, see pages 80 and 81.
Distance:	Approx. 70 kilometres, Kingsvale to Boston Bar.
Travel Time:	Three to four hours.
Condition:	Some steep, rough gravel sections, closed in winter.
Season:	June through October.
Topo Maps:	Lytton 92 I/SW (1:100,000). Merritt 92 I/SE (1:100,000) Tulameen 92 H/NE (1:100,000).
Forest Maps:	Merritt Forest District Recreation Map.
Communities:	Merritt and Boston Bar.

There is an historic "short" cut through the North Cascade Mountains that connects the Coldwater Road, north of Kingsvale, with Trans Canada Highway 1 at Boston Bar. Interior Salish Indians from the lower Fraser Canyon used this route while trading with Natives of the Nicola. For a couple of seasons in the late 1840s, it heard the creak of Hudson's Bay Company pack saddles as furs and supplies were transferred to and from Pacific ports. One of the valleys along the route was also the site of the Nicola Valley's only logging railway.

Today, this is a summer route through a mountain pass that's close to 1350 metres in altitude. The inviting features about this high-country route include some excellent lake and stream fishing, (complete with Forest Service recreation sites), huckleberries near the summit, hunting opportunities and, of course, the opportunity to get totally lost in the back country. Negatives include some steep sections and, depending on which eastern approach you take, some potentially muddy sections and, as I mentioned, the opportunity to get lost in the back country.

The term "short" needs to be qualified. At about 75 kilometres from Exit 256 on the Coquihalla Highway to Boston Bar, it is definitely shorter than the north route via Merritt and Spences Bridge or the south route via Hope. However, if time or the family sedan are any consideration, you may be wiser to stick to blacktop.

From Coquihalla Highway 5 Exit 256, follow the Coldwater Road 13 kilometres north to Patchett Road. If you are coming from Merritt, Patchett Road leads west off the Coldwater Road, 18 kilometres south of Highway 5A. With the junction of Patchett Road and Coldwater Road as kilometre 0, head west across the old KVR right-of-way. Continue past a few old log buildings and across Coldwater River at km 0.9.

Follow the main road as it swings to the north and over a knoll, away from the main east-west Midday Valley. Part of this area is Indian Reserve and it was also the route of a 1920's logging railway. A major forest fire ravaged the area in July, 1960. The scars of the Dean Fire are still visible on the mountainsides.

Patchett Road draws a fairly wide arc and then by km 4.0 it heads westward. The route here is well-maintained gravel. If you are interested in wildflowers, you'll see arrowleaf balsamroot in May followed by at least three varieties of arnica in June.

There are side roads north at km 5.7 and 6.7 — keep left on the main road through the valley.

Patchett Road becomes the Spius Creek Forest Service Road just past the 10K marker and then it begins a southwest route along the canyon edge high above Spius Creek. The road climbs steadily affording some excellent views of the canyon and surrounding mountains near the 14K marker. Keep right at the junction marked Airport Road — the airport was built to fight the Dean Fire.

After clinging to the canyon wall for seven kilometres, the road descends to a one-lane bridge across Spius Creek at km 21. Follow the creek upstream past some interesting pools and a few convenient spots to park your vehicle while you check them out. Follow the main road southwest (straight ahead) past a well-used side road to the right near km 25. Then, near 26K, begin a switchback climb out of the valley. Keep straight ahead (left) at junctions near the 27K and 29K markers and by km 32 you should see the 500 kV power lines. After passing under the lines, watch for a road to the left that goes down a short grade. The road straight ahead continues around the mountainside, following the hydro right-of-way to Uztlius Creek, but it may be impassable.

Head left across a small creek before joining with the powerline road from Gillis Lake and Kingsvale. Reset your trip meter at the junction, and keep right here as well as at the next junction where a newer road climbs up the hill. Then follow the older road south and then southwest as it climbs through a stand of timber. You should pass the 33K marker at about km 0.7 and then break out into an old logged area near km 2.9. Keep left and up the hill at a junction near km 3.2. At km 4.2, you should be south of a meadow that is part of the headwaters of Uztlius Creek.

You have now crossed over into the Anderson River and Fraser River drainage area at an elevation of about 1310 metres.

Figure 19
Repairing the Uztlius Creek bridge at 10K. (Photo CLD-163)

The BC government 1939 Hope-Princeton map shows the 1848-49 Hudson's Bay Company (HBC) Brigade Trail through this area. It started at the Fraser River near the present Alexandra Bridge and climbed up the mountainside behind Alexandra Lodge. Bob Harris, in *The Best of B.C.'s Hiking Trails*, details the lower end of the trail, but the sections through the Uztlius Creek canyon, east to Gillis Lake and northeast to the Coldwater River near the mouth of Midday Creek, have long been lost to forest fires and logging. As you will see when you drive through the area, this is steep, inhospitable terrain, and after two seasons, HBC Chief Trader A.C. Anderson and Henry Peers found a slightly better route to the east of the Coquihalla Canyon.

Keep right at junctions at km 4.6 and 5.1, left at km 5.7 and right at

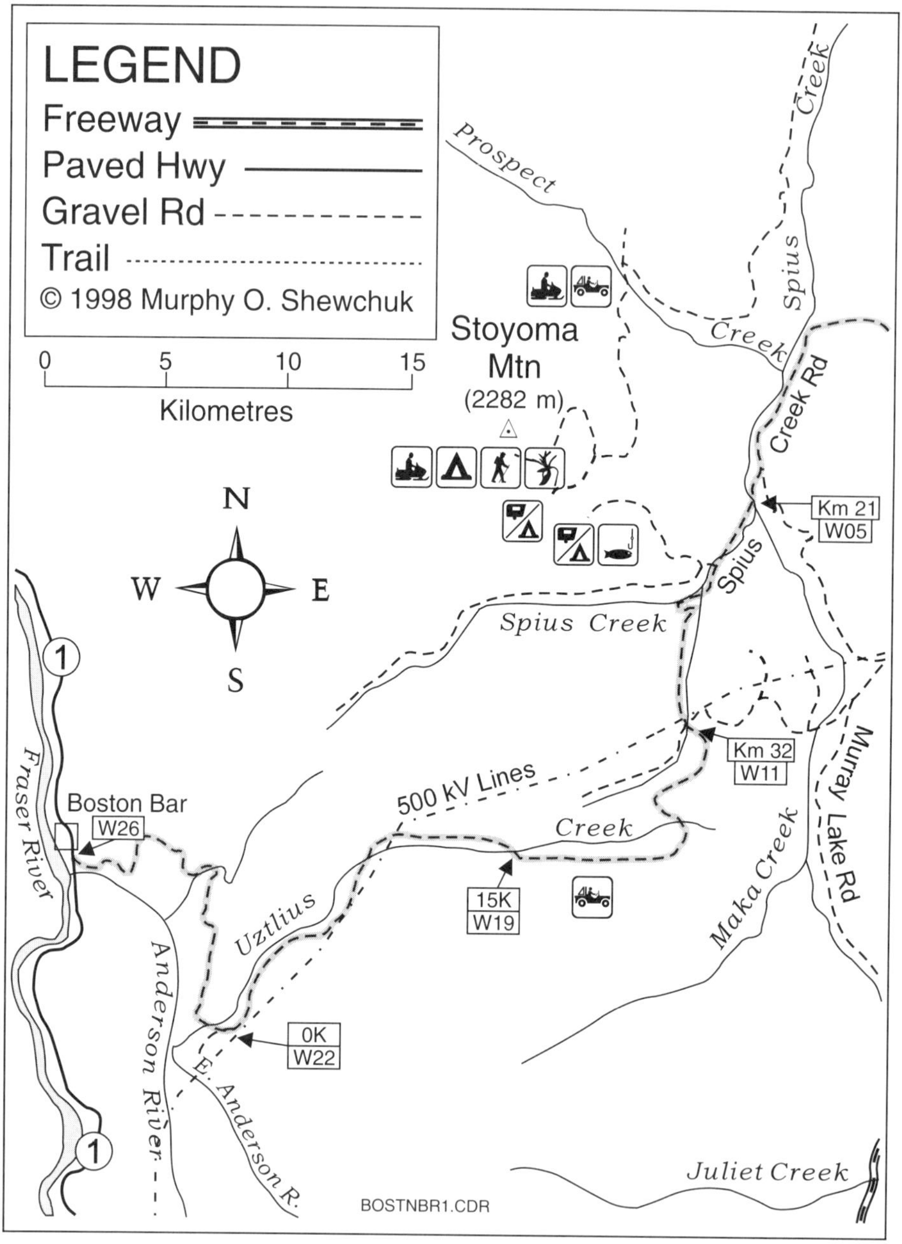

Map 14
Backroad to Boston Bar — West Section

6.1 (near the 22K marker on the countdown toward Boston Bar) and again at km 6.7. Side roads come in from the right (north) near km 7.5 and 8.6, but by now you are on the downhill run on the Uztlius Creek Forest Service Road. You should see a steady decrease in the roadside

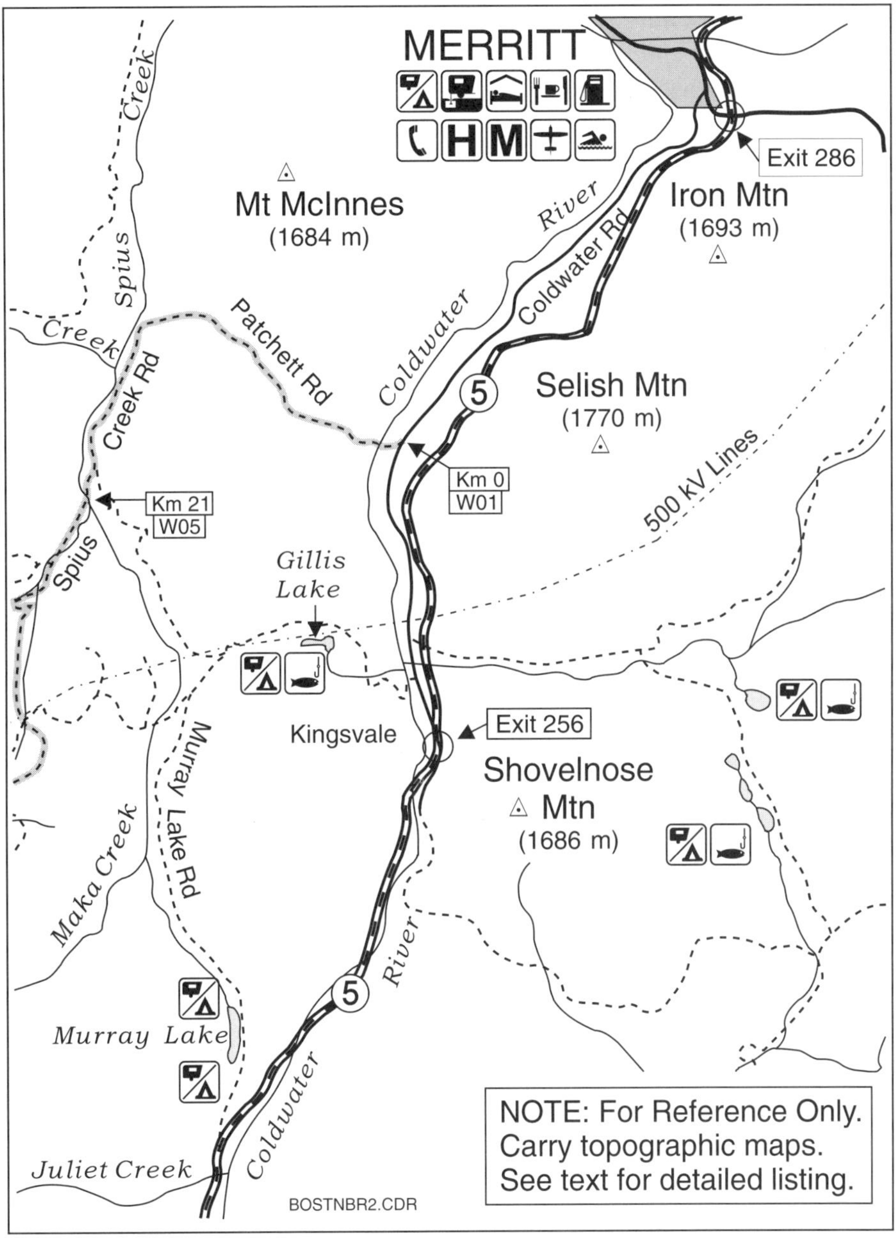

Map 15
Backroad to Boston Bar — East Section

marker numbers — if they start to increase, you've made a wrong turn.

You'll have some steep grades to descend near km 12, but by km 15 you'll be into the good road with numerous places to pull over and enjoy the scenery or try a little stream fishing. There are numerous crossings of Uztlius Creek and nearly as many side roads to explore as you make the descent. If you're interested in camping, there's a Forest Service recreation site near km 28.5.

Less than a kilometre later you should be at the 0K junction of Uztlius Mainline (which you just came down) and Uztlius-East Anderson River Road. Some of the names can be a little confusing, particularly if you're trying to decide where you've been, but swing right (north) and continue along the good gravel road through the timber. You'll cling to the hillside for the next 10 kilometres, with some excellent views of the Fraser Canyon to the north.

Then it's time to check your brakes again for the steep, winding descent to the Trans Canada Highway near the Anderson River bridge, just two kilometres south of Boston Bar.

If you have a GPS receiver in your vehicle, the waypoints below should help you get through the maze of junctions and side roads.

•••

GPS References for major points of interest. See page 9 for details.

Wpt	Description	Latitude	Longitude
W01	Jct. Coldwater - Patchett Road	N49° 59' 10.879"	W120° 55' 04.416"
W02	Coldwater River bridge	N49° 59' 01.493"	W120° 55' 42.383"
W03	10K Spius Cr. FS Road	N50° 01' 29.944"	W121° 00' 56.839"
W04	Jct Airport Road	N49° 58' 38.261"	W121° 04' 36.511"
W05	Spius Creek bridge	N49° 58' 02.148"	W121° 04' 50.300"
W06	2nd Spius Cr. bridge	N49° 56' 30.185"	W121° 06' 12.646"
W07	25K Marker - Spius Cr. FS Rd	N49° 56' 11.626"	W121° 06' 31.746"
W08	Jct. Spius Cr.-Spius Cr. S. Rd	N49° 56' 07.725"	W121° 06' 41.537"
W09	Jct. keep left	N49° 55' 40.341"	W121° 06' 58.995"
W10	Jct. keep left	N49° 54' 32.498"	W121° 07' 05.715"
W11	Jct. go left near power line	N49° 53' 10.268"	W121° 06' 53.008"
W12	Jct. keep right	N49° 51' 46.280"	W121° 06' 56.233"
W13	Jct. keep left	N49° 51' 17.583"	W121° 07' 30.531"
W14	Jct. keep right	N49° 51' 14.879"	W121° 07' 55.405"
W15	20K Marker	N49° 51' 14.068"	W121° 09' 02.359"
W16	Old Uztlius FS Rd from right	N49° 51' 13.393"	W121° 09' 17.519"
W17	Jct. go right	N49° 51' 08.545"	W121° 09' 36.561"
W18	Jct. keep right	N49° 51' 06.228"	W121° 11' 46.646"
W19	15K Marker near bridge	N49° 51' 14.146"	W121° 12' 19.399"
W20	10K Marker near bridge	N49° 51' 07.715"	W121° 16' 08.380"
W21	Rough FS campsite	N49° 47' 37.254"	W121° 20' 45.274"
W22	0K Jct. E Anderson FS Rd	N49° 47' 26.169"	W121° 21' 10.514"
W23	10K Marker	N49° 49' 37.837"	W121° 21' 53.059"
W24	Jct. Uztlius-Stoyoma FS Rd	N49° 50' 55.239"	W121° 22' 41.802"
W25	5K Marker - start descent	N49° 51' 01.747"	W121° 23' 20.059"
W26	0K Jct. Hwy 1 Boston Bar	N49° 50' 58.098"	W121° 26' 00.580"

•••

19

Kane Valley Road

Statistics	For maps, see pages 85 and 91.
Distance:	35 km, Coquihalla Highway 5 to Highway 5A/97C.
Travel Time:	One to two hours in dry weather.
Condition:	Gravel road with rough sections.
Season:	Anytime except spring break-up.
Topo Maps:	Merritt 92 I/2 (1:50,000). Aspen Grove 92 H/15 (1:50,000).
Forest Maps:	Merritt Forest District Recreation Map.
Communities:	Merritt.

Kane Valley Road is a back country thoroughfare, less than three hours from the Lower Mainland, that offers three-season enjoyment for almost any outdoor interest.

The three-season part deserves a bit more elaboration. Unlike the "dry weather only" recommendation I usually hang on the more remote backroads, this route really should be avoided only during late March and April, a season known as *spring break-up*.

Those of you who aren't familiar with the term "spring break-up" as it applies to this part of the B.C. Interior should think of it as the time of the year when everything turns to mush. The ice breaks-up on the lakes, turning the frozen surface to slush and trapping a few unwary snowmobilers and ice fishermen each year. The gravel roads break-up, revealing a base of bottomless red clay mixed with a few oilpan-eating boulders. Cross-country skis break-up when their owners collide with a dislodged boulder or patch of bare ground. Drivers have been known to suffer mental "breakdown" when they discover that they are stuck up to their axles in the middle of an otherwise normal road, but that's another subject.

Spring excluded, there are three seasons and several reasons to explore this east-west valley in the mountains south of Merritt. My wife, the cross-country skiing fanatic, will obviously tell you that the best season is winter and the best reason is the 45 kilometres of well-marked cross-country ski trails. She should know, because she helped promote them way back when cross-country skiers used wooden skis and bamboo poles. I certainly won't argue with her, I have enough trouble keeping up without trying to talk and breathe at the same time.

Figure 20:
Ducks are a frequent sight on the Kane lakes chain. (Photo 219-D-19A)

My neighbor, the fisherman, thinks the early summer and fall seasons are the best. Mind you, he's like me in not being able to understand people who sit by a hole in the ice and try to entice a trout to swallow a kernel of corn on the end of a fishing line. Hot rum and a crackling fireplace seem to be better surroundings if one wants to sit still in winter.

The Kane Valley Road has two ends, unlike some of the roads we have covered *and* it has the same name at both ends (ditto). Most cross-country skiers approach it from the east off what is now Highway 5A/97C, 18 kilometres south of Merritt and two kilometres north of the world famous Corbett Lake Country Inn. (Please excuse this blatant commercialism, but owner Peter McVey is a chef whose cooking, fly-tying and rod-making skills ARE world-famous.)

Because we both know that the odds are that you will be driving up the Coquihalla Highway from Hope and that you will have a four-wheel drive vehicle or a decent set of tire chains, I will outline your route from the west end, near the former Kettle Valley Railway (KVR) community of Kingsvale. As the Kane Valley Road isn't first on the snow plow priority list, snow tires and chains aren't an idle suggestion.

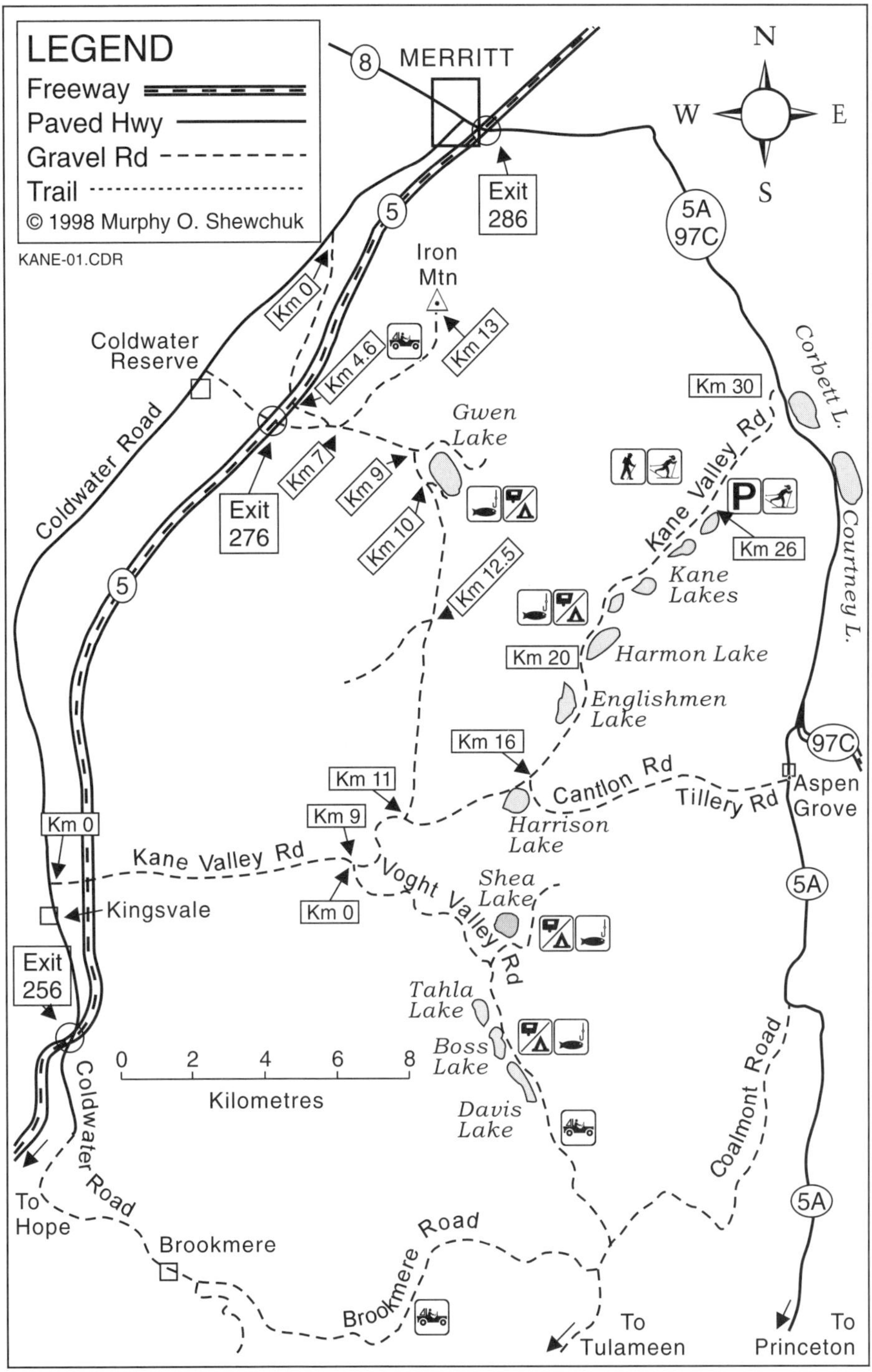

Map 16:
Iron Mountain — Kane Valley — Voght Valley area.

To get to the west end of Kane Valley Road, take Exit 256 off Coquihalla Highway 5, approximately 80 kilometres north of Hope or 30 kilometres south of Merritt. Continue north down the Coldwater River valley (south will get you to Brookmere and a dead end in winter) through a former railway underpass at Kingsvale and up the hill to the junction of Coldwater Road and Kane Valley Road, about 3.5 kilometres from the freeway.

Swing right at the junction (km 0) and follow the road as it climbs steadily eastward up Voght Creek, passing under the freeway at km 1, and then through scrub pine and aspen.

Voght deserves more than a passing mention. William Henry Voght was the founder of the City of Merritt. He was born in 1832 in Holstein, now part of West Germany, and came to North America as a teenager. After temporary diversions in Iowa, New Orleans and the California Gold Rush, in 1858 he joined a group headed for the Fraser River Gold Rush. After two seasons of chasing the elusive yellow metal, young Bill Voght took up farming across the Fraser River from Boston Bar. He farmed there until 1873 before moving to where Merritt stands today. When coal was discovered in the nearby hills, Voght promoted the construction of a railway into the valley and surveyed the new coal mining community on part of his farm. After a short illness, Voght died in February, 1911, less than two months before the City of Merritt received its charter. Voght Valley was named after Voght's eldest son, also named William, who settled here in the late 1880s.

The Kane Valley Road, gravel, twisting and narrow in places, levels off somewhat near km 5 and parallels the creek bed through some excellent moose pasture. Merritt's Tolko Industries is selective logging throughout much of the valley and its efforts are clearly visible here.

A lone log house in the field to the south of the road near km 8 is believed to have been built by John Smith when he settled in the valley in the late 1880s. His wife, Jessie Ann, details the difficulties of their homesteading life in *Widow Smith of Spence's Bridge* (Sonotek Publishing Ltd., 1989). They lived here for 10 years before moving to Spences Bridge in 1897.

Approximately one kilometre farther along, a sign marks the road to Boss, Davis, Tahla and Shea lakes. Known as Voght Valley Road, this 15-kilometre-long backroad was once part of the trail from the Nicola Valley to the mines at Granite Creek. Today, recreationists find Voght Valley of particular interest for the lakes and the Forest Service recreation sites at each one. Voght Valley Road joins the Coalmont Road about 20 kilometres south of Aspen Grove.

Figure 21
Kane lakes trout. (Photo NVG-212)

Keep straight ahead at the junction as the Kane Valley Road passes through Voght Valley Ranch (now part of Nicola Ranch) and begins to swing northeast up Kanevale Creek. Cantlon Road, another side road just past man-made Harrison Lake, winds 10 kilometres easterly through the upland timber to Highway 5A, coming out as Tillery Road just south of Aspen Grove. It is a pleasant drive in the autumn, but is rough and narrow in places.

As much of the log hauling is done to Highway 5A, roadside numbers begin a countdown near Voght Valley Ranch. The Cantlon/Tillery Road junction is about 16 kilometres from Kingsvale at the 14K marker. Englishmen Lake is the next point-of-interest along the route. Like most lakes in the valley, it has been dammed for agriculture irrigation.

Harmon Lake at km 20 [10K] is the best known and busiest of the Kane Lakes chain. A large open Forest Service recreation site here is seldom empty except in winter when the only access is over the snow. An interpretive trail, with a series of information plaques, winds through the forest and grassland on the southeast side of Harmon Lake.

Harmon Lake marks the west end of the Kane Valley cross-country ski trail system. This network extends to the ridges on both sides of the valley and along the valley to the 2K signpost, with several maintained parking lots and rest areas.

A sign marks Lower Kane Lake near km 21. At the time of writing this lake was a catch and release fishery with a gasoline motor prohibition. Before the restrictions went into effect, I watched an excited lady pull in a rainbow trout that must have weighed two kilograms after half-a-dozen casts with light fly-fishing tackle. Upper Kane Lake is a few minutes farther along the road.

A ski trail named "Overeasy" (a blatant lie) near the junction of

Upper and Lower Second Lakes marks one of the many ski trails that wind up to the ridge on the north side of the valley. (See the map on page 91 for details.) With the help of local volunteers, Forest Renewal BC programs and the Forest Service, the extensive network of trails on both sides of the valley from Harmon Lake to Highway 5A has been well-maintained during the past several winters.

Thirty-kilometre-long Kane Valley Road climbs steadily from Kingsvale, reaching an elevation of about 1220 metres before dropping quickly in the last two kilometres to Highway 5A near Corbett Lake.

If you are looking for some mid-winter backroads exploring, give Kane Valley Road a try. But don't forget your skis *and* your tire chains. If you must wait for summer, bring your boat, your hiking boots or your mountain bicycle — and your binoculars — the trails can be particularly interesting for naturalists and bird watchers.

Figure 22
Hiking the trails. (Photo HIK-037)

•••

20

Kane Valley Trails

Statistics	For maps, see pages 85 & 91.
Distance:	Approximately 45 km of marked trails.
Travel Time:	Not applicable.
Condition:	Trails are marked and maintained regularly.
Season:	Mid-November to mid-March.
Topo Maps:	Merritt 92 I/2 (1:50,000). Aspen Grove 92 H/15 (1:50,000).
Forest Maps:	Merritt Forest District Recreation Map.
Communities:	Merritt.

When snow covers the hills in a white blanket, the Kane Valley, south of Merritt, becomes a haven for outdoor enthusiasts. The chain of lakes along the valley floor attract ice fishermen (and women). A network of trails and old roads on both sides of the valley become the playground of Nordic skiers from the Nicola Valley as well as dry snow enthusiasts from the lower mainland.

The Kane Valley, twenty minutes south of Merritt, an hour west of Peachland or an hour north of Princeton, is only a few minutes off Highway 5A/97C. With a base elevation of 1160 metres the timbered valley has a micro-climate that can provide ideal cross-country ski conditions. The snow is generally dry, although seldom more than 1 to 1.5 metres deep, and the air temperature is usually between 0 and -15 degrees Celsius throughout the winter.

The better winter access is via Kane Valley Road from Highway 5A/97C near Corbett Lake, 18 kilometres south of Merritt. The Ministry of Highways maintains several parking areas at points where the ski trails cross the road. The first is approximately two kilometres from Highway 5A/97C. The third parking lot (P3 on the map), near the 4K marker, is larger and has a shelter and room to park a bus. (Parking lot P2 on the Kane Valley Trails map has not been kept plowed because of difficult winter access.) The three remaining parking areas are smaller, but provide good access to the trails. Access is also available from the west end of Kane Valley Road, near Kingsvale and Exit 256 on Coquihalla Highway 5 (see page 83).

Trail conditions vary, depending on snowfall. Maintenance of the 45 kilometre trail network is subject to the availability of funding. After nearly two decades of ad-hoc ski development, the Nicola Nordic Ski Club is making every effort to keep the trails groomed.

Figure 23
Late afternoon on the Skyhigh Trail near the top of Ravine.

If you are out for an easy ski with young or new skiers, try the Kane Trail. It begins at parking lot P3 and follows an old logging road along the south side of the valley. The trail climbs easily for several kilometres, with a few twists and dips to add some challenge, before winding down 1.5 kilometres to a shelter at January Landing.

The trail ends at Kane Valley Road at a footbridge and dam at the west end of Lower Second Lake. You can return by the same route or cross the road and return via Overeasy, Gulch and Vale on the north side of the valley.

The much more challenging Harmon Trail is also accessible from parking lot P3 for a full loop that totals about 16 kilometres. It climbs up Vale initially, and then continues to climb up the north side of the valley as the Gulch Trail. It joins the Harmon Trail at the foot of the "gulch". After a winding climb to the summit, it makes a long run down an old logging road to Harmon Lake.

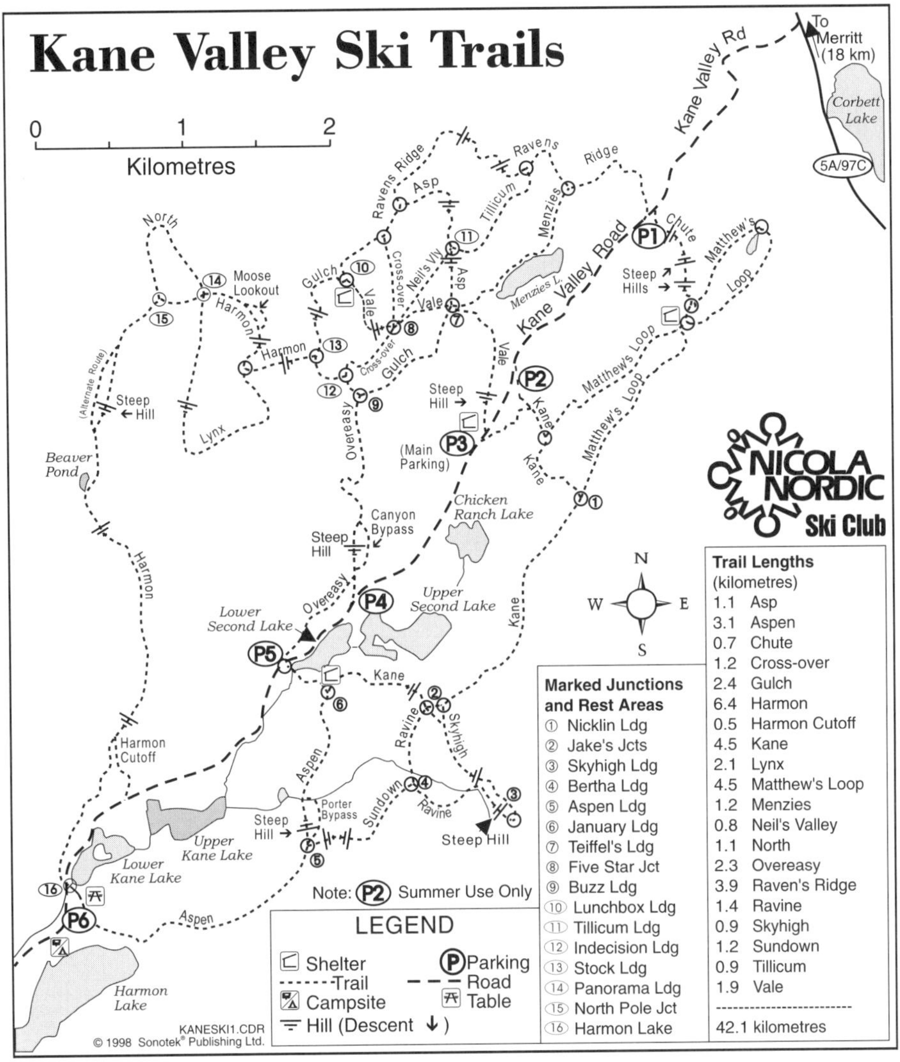

Map 17
Kane Valley Nordic Ski Trails

You can cross Kane Valley Road here and after skirting Harmon Lake, climb east and up the south side of the valley on the Aspen Trail. The Sundown and Ravine trails will allow you to keep high on the south slopes of the valley, or you can follow Aspen down to January Landing where there is often a campfire going.

The section of trail from the Sundown junction to January Landing has a long, steep hill that will give you a thrill or a chill, depending on your skiing abilities. If you aren't equipped for narrow, steep trails, an alternate route was recently cut through the timber in this area. From

January Landing, you can follow the Kane Trail back to the second parking lot.

Some sections of this route are fairly steep and will require good climbing and descending skills, although they wouldn't qualify as expert telemark slopes.

If you plan to make a weekend of it, accommodations in the Merritt area are plentiful, although you may require reservations. Tourism British Columbia's accommodation guide provides a full listing. You can also contact the Merritt and District Chamber of Commerce for more information by phone at (250) 378-5634 or fax at (250) 378-6561.

The Forest Service and the ski club have posted up-to-date maps at trail register boards near the parking lots. Map/brochures are also available in advance by sending a stamped, self-addressed #10 envelope to the Nicola Nordic Ski Club, Box 1499, Merritt, BC V1K 1B8.

These maps provide detailed information on the Kane Valley Trails, including length, level of difficulty and location of shelters. The club will also send you information about their membership package.

If cross-country skiing doesn't fit into your lifestyle, the half-dozen lakes in the area attract ice fishermen as well. The Kane Valley also has a life when the snow has gone. Summer activities on the trail system include hiking and mountain bicycling as well as bird-watching and wildlife viewing. The Forest Service recreation site at Harmon Lake is one of the busiest in the area. However, despite the activity, the rolling grassland and light timber creates a sense of privacy not often found in campgrounds. The Forest Service also has an excellent interpretive forest on the south shore of Harmon Lake.

As a final reminder, drive with caution winter and summer as loggers and ranchers also use the main road and many of the side roads.

•••

21

Gwen Lake Road

Statistics	**For map, see page 85.**
Distance:	14 kilometres, Coldwater Road to Iron Mtn Summit. 11 kilometres, Coldwater Road to Gwen Lake.
Travel Time:	One half hour.
Condition:	Gravel, some rough sections.
Season:	May through October.
Topo Maps:	Merritt 92 I/2 (1:50,000). Aspen Grove 92 H/15 (1:50,000).
Forest Maps:	Merritt Forest District Recreation Map.
Communities:	Merritt.

Iron Mountain was the site of considerable mining exploration in the late 1800s and early 1900s. George M. Dawson, the noted geologist, first used the Iron Mountain name in his Geological Survey of Canada Report of 1877-78. The BC Mines report for 1896 indicates that twenty-four mineral claims had been staked on Iron Mountain that year. In addition to iron, also present were gold, silver and copper, but little came of the prospects.

In 1927, Emmett Todd, a veteran prospector, discovered a silver-lead vein one kilometre south of the summit that showed considerable promise. A shaft was sunk more that 30 metres deep on the Leadville claim before a fault was encountered and the vein was lost. In 1947, the shaft was reopened as the Lucky Todd Mine and 32 tonnes of ore were shipped to the Trail smelter, containing 2000 grams of silver, 5360 kilograms of lead, and 220 kilograms of zinc.

The effort was shortlived, however, and except for prospectors and rockhounds searching the old waste dumps for traces of jasper, peacock copper and malachite, Iron Mountain has remained quiet.

Gwen Lake Road (Veale Road at this point) swings east from the Coldwater Road five kilometres south of Highway 5A/97C. With the junction as km 0, follow Veale Road up the hillside through hay fields at km 0.9 and 2.2 and under the Coquihalla Highway at km 4.6. Keep left at the Comstock Road junction at km 6.2.

Comstock Road joins Coquihalla Highway 5 at Exit 276, about 1.5 kilometres to the west (down the hill). It then continues down the mountain to the Coldwater Indian Reserve village.

A junction seven kilometres from the Coldwater Road marks the turnoff to Iron Mountain, with the summit six kilometres to the north and Gwen Lake four kilometres to the east. At the 6.5K marker, approximately four kilometres up Iron Mountain, a very rough road leads to the west. It winds through the open timber for about 1.3 kilometres before reaching the site of an old mine shaft. You may find some interesting ore samples near the shaft and where the mountain has been trenched.

Iron Mountain is a maze of communications buildings, but despite this, the 1693-metre peak offers an excellent view of Sugarloaf Mountain and the Nicola Valley grasslands.

Back at the junction at km 7.0, the road to the southeast continues toward Gwen Lake. The first road to the left provides access to an Indian Reserve. The one kilometre access road to the Gwen Lake Forest Service recreation site branches to the left near km 10. This road is very rough with boulders in the mud holes, and may best be explored on foot.

Beyond the Gwen Lake junction, the main road continues through the uplands, eventually following Howarth Creek down to the Kane Valley Road. The section between Gwen Lake and the Kane Valley Road is a maze that should not be attempted without a good topographic map, a compass, a full fuel tank and plenty of time. If you do decide to try it, keep left at a junction near km 2.5 and follow the road across the 500 kV powerline at km 3.7. Keep right at a junction in a clearing near km 5 and again at an old junction near km 7.7. If you make all the right moves, you should reach Kane Valley Road about 11 kilometres from the Gwen Lake junction and 11 kilometres from the Coldwater Road.

•••

GPS References for major points of interest. See page 9 for details.

Wpt	Description	Latitude	Longitude
W01	Veale Road and Coldwater Road	N50° 03' 54.536"	W120° 48' 00.220"
W02	Coquihalla Highway underpass	N50° 01' 50.280"	W120° 48' 57.510"
W03	Jct. Iron Mountain to left	N50° 01' 03.116"	W120° 47' 43.490"
W04	Jct. old mine workings to left	N50° 02' 01.513"	W120° 45' 49.233"
W05	Iron Mountain Summit	N50° 02' 46.802"	W120° 45' 18.234"
W06	Jct. Gwen Lake to left	N50° 00' 14.134"	W120° 45' 58.030"
W07	Gwen Lake Rec Site	N50° 00' 08.745"	W120° 45' 26.626"

•••

22

Okanagan Connector

Highway 97C

Statistics	For maps, see pages 97, 100, 106 & 160.
Distance:	90 kilometres, Aspen Grove to Westbank. 115 kilometres, Merritt to Westbank.
Travel Time:	Approximately one hour.
Condition:	Four-lane, divided highway except for a two-lane section north of Aspen Grove. Limited access.
Topo Maps:	Ashcroft 92 I (1:250,000). Hope 92 H (1:250,000). Penticton 82 E (1:250,000).
Forest Maps:	Merritt Forest District Recreation Map. Penticton Forest District Recreation Map.
Communities:	Merritt, Aspen Grove, Westbank & Peachland.

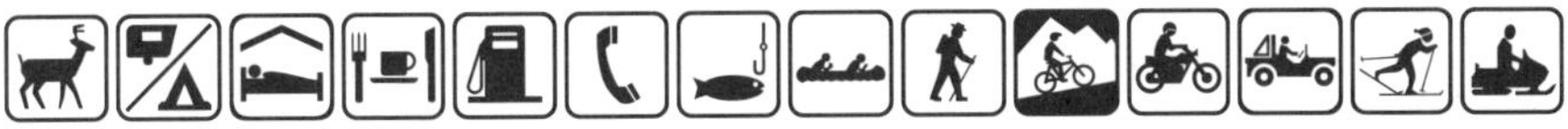

Wildlife made the first trails across the plateau country separating the Nicola Valley from Okanagan Lake. They moved steadily, driven by the seasons, their need for forest cover while rearing their young, and the never ending search for food. Natives followed these trails, both to find the animals and to trade with other native groups living in the Similkameen, Nicola and Thompson valleys.

David Stuart, of the Pacific Fur Company, headed the first white party to travel the trails through the Okanagan Valley. In September, 1811, his party of fur traders, assisted by friendly Native guides, travelled as far north as Kamloops. They quickly established the plateau trail network as part of their trade route from the Interior to the Pacific Ocean at the mouth of the Columbia River.

In 1859, as the fur trade waned and the Fraser River gold rush flourished, the plateau trails began to feel the traffic of miners. Initially, their destination was the bars of the Fraser, but Tranquille Creek, near Kamloops, and Granite Creek, northwest of Princeton, soon attracted their attention. Then, as interest shifted to the Boundary District, the gold rush traffic moved in the opposite direction. Rock Creek became a destination of choice.

Ranchers, loggers and prospectors turned the horse trails into wagon roads. These, in turn, were slightly improved by crews building high voltage electric lines. But, from a recreational perspective, the plateau route from the Nicola Valley to the Okanagan Valley remained the domain of fisherman, hunter and backroad explorer.

The idea of a highway across the plateau is not a new one. Earlier this century, members of the former Kelowna Board of Trade, now the Kelowna Chamber of Commerce, began to promote this alternate route to Merritt. For many years, they organized a cavalcade of vehicles from Kelowna to Merritt during the early summer months. Until the opening of the Okanagan Connector (Highway 97C) on October 1, 1990, determination and a pick-up truck were prerequisites for a successful cross-country trip from Merritt to Kelowna.

According to the B.C. Ministry of Transportation and Highways, the 108 kilometre long route connecting Coquihalla Highway 5 at Merritt with Highway 97, between Peachland and Westbank, was completed one year ahead of schedule and at the budgeted cost of $225 million. The Okanagan Connector cut 1.5 hours off the usual highway travelling time between the Lower Mainland and the Central Okanagan. It also reduced the travel time between Merritt and Kelowna by half.

From an outdoor recreation perspective, the Okanagan Connector (Highway 97C) has its greatest value as a "connector". It improves access to the Okanagan's five ski hills for Nicola Valley and Lower Mainland residents. It also reduces travel time for Okanagan fishermen heading for the lakes surrounding Merritt. However, the highway was deliberately designed to serve as a throughway, with little access to the many fine lakes of the region.

The reasons behind this limited access, according to government sources, are several. Uppermost in mind was the inability of the fragile upland lakes to withstand the onslaught of tens of thousands of new visitors. Pennask Lake, the source of a large portion of British Columbia's Rainbow trout hatchery eggs, is too close to the highway for comfort. Over-fishing at Pennask Lake would have detrimental effects on fishing throughout the Interior. The numerous small lakes in the Paradise Lake area would also suffer a similar fate.

Ranching interests, particularly the Douglas Lake Cattle Company, are also opposed to wide open access to the plateau. Their concerns

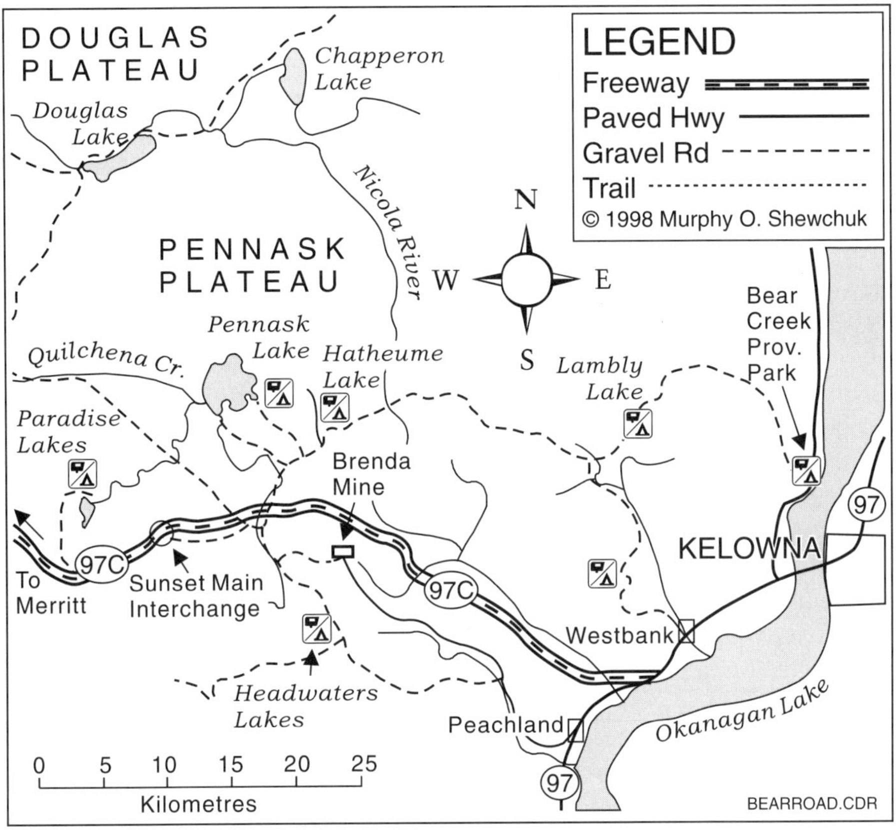

Map 18
The Okanagan Connector: Paradise Lakes to Kelowna.

included loss of valuable rangeland to highway right-of-way and increased cattle harassment.

A detailed description of the route serves little purpose, but here are a few highlights. With the junction of Highway 5A and 97C as the km 0 reference, you will get a clear view of Alleyne Lake as you descend the hill near km 3.2. (See *Kentucky-Alleyne Provincial Park*, page 123.) Although not easily seen when you travel east, Pothole Lake is visible on the north side of the road near km 7.2.

The Loon Lake Road exit at km 16.0 can take you south into a land of lodgepole pine and upland meadows. The gravel backroad also provides an alternate route past Loon Lake to Kentucky-Alleyne Provincial Park and Highway 5A, south of Aspen Grove.

In addition to the nearby Elkhart Lodge, the Elkhart Road exit at km 30 offers access into the Paradise Lakes area north of the Connector. Here a network of gravel roads and trails lead to Paradise Lake Resort and several Forest Service recreation sites.

Sunset Main Road, near km 42, also provides access to the plateau country. There is a 6.7-kilometre-long rough gravel road along the north slopes of Pennask Mountain that links the interchange with the Bear Forest Service Road. You can follow this backroad northeast and down to Okanagan Lake near Bear Creek Provincial Park. Sunset Main Road continues southeast into the Headwaters Lakes area before descending to Peachland. (See *Pennask Lake Road* on page 157.)

The Okanagan Connector crests near km 54 (about 1728 metres above sea level) and begins the 32 kilometre schuss to the Okanagan. The highway skirts the Trepanier Creek canyon, passing the Brenda Mine tailing pile near km 63.

Brenda Mine, a copper/molybdenum operation, closed in 1990 after two decades of production. A quartz vein on the site was originally worked by the Sandbergs of Kelowna in the late 1930s and early 1940s as the "Copper King" property. It was then abandoned until re-discovered in 1954 by Bob Bechtel, a Penticton prospector. It took ten years of effort before Brenda Mines Ltd. was formed and another five years and $62.5 million before the mine began production in 1970. A member of the Noranda Group, Brenda produced 278,000 tonnes of copper, 66,000 tonnes of molybdenum, 125 tonnes of silver and two tonnes of gold.

There is an ungulate overpass about seven kilometres from the bottom of the hill that is said to be the first of its kind in Canada.

The tourist information centre a few kilometres from the junction with Highway 97 provides the last stop before reaching the two-lane traffic corridor between Peachland and Westbank.

•••

23

Paradise Lakes Area

Statistics	For map, see page 100.
Distance:	7.5 kilometres, Elkhart Exit to Paradise Lake Resort.
Travel Time:	15 minutes.
Condition:	Mostly good gravel road.
Season:	Maintained year around.
Topo Maps:	Paradise Lake, BC 92 H/16 (1:50,000)
Forest Maps:	Merritt Forest District Recreation Map.
Communities:	Merritt, Peachland and Westbank.

Gordon Mountford, born and raised in Summerland, first saw the upland country around Paradise Lake after World War II. "When I got out of the army in 1945, I spent a month exploring the area with a horse. I stopped at Teepee Lake, Siwash Lake and all the lakes up in there. I didn't care if I ever went back to civilization."

He did go back to civilization, but the lure of the high country was too much to resist. "In the spring of 1949, we opened the Paradise Lake Fishing Camp. There was absolutely nothing when we got there."

While there may not have been anything at Paradise Lake when Gordon Mountford arrived, he certainly hasn't laid claim to discovering the lake. The Natives knew the lake as Hutkaneta or Hutkanetco, which Mountford believes meant "pretty at the end." "Paradise", was tacked on by Oscar Schmidt, who trapped in the area around 1903-1910.

One of Paradise Lake's attractions must have been the isolation. For over 40 years, the road to the fishing camp was barely more than a cart track. But it didn't keep serious campers and fishers away.

Gordon Mountford sold the Paradise Lake Fishing Camp to the Grassers in 1962. Ken Collins bought it in the fall of 1964, then Hugh McGowan Sr. bought the Paradise Lake operation in 1966 as a retirement venture. When he died in 1973, his wife, Ruby, continued to operate the fishing camp until she, too, passed away.

The Paradise Lake Co-operative Association bought the property in

1992. Since then, a new lodge has been built as well as approximately 20 privately owned cabins, many of which are available for rent.

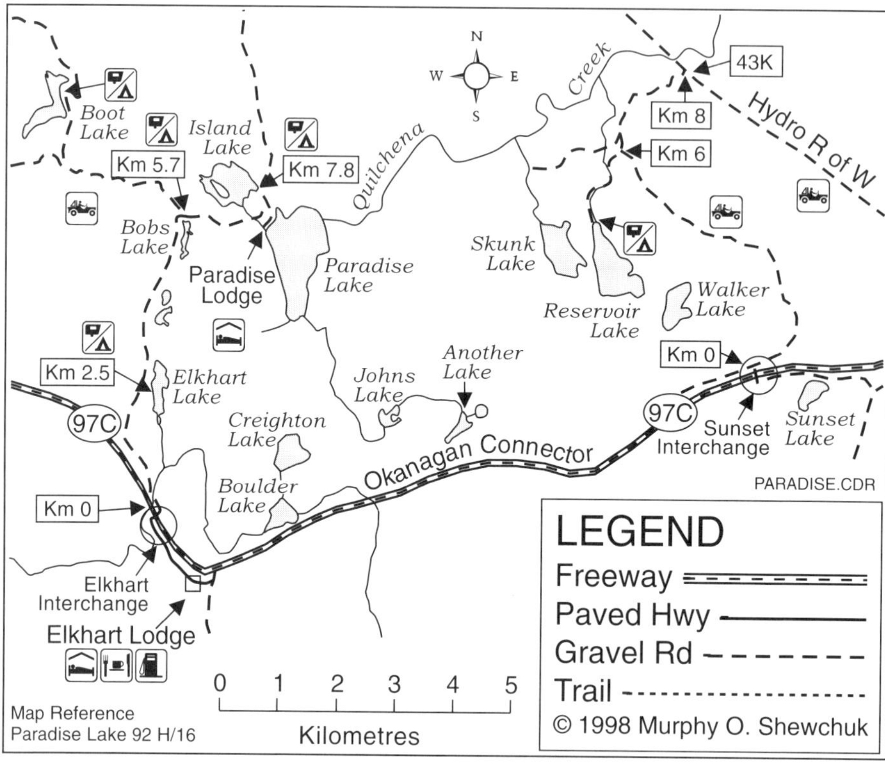

Map 19
Paradise Lake area.

Paradise Lake has come up in the world. No longer is it a " fishing camp" with rustic cabins. It is now a " resort" with a new log lodge, private rooms and three meals a day. But beyond the natural-finish logs and sheet-metal roofs are a network of old roads and trails that can still lead you to your own private fishing hole.

The simplest way there is to exit Highway 97C at the Elkhart Interchange (km 0.0) and follow the gravel road northwest across Elkhart Creek. (You may want to top up your fuel tank a Elkhart Lodge first.) After about a kilometre, the road joins an old tote road that was used during highway construction. Watch for a rough road to the east (right) at km 2.5. If you've found the right road and you're concerned about your muffler, it's only a few minutes walk to the Forest Service recreation site on one-kilometre-long Elkhart Lake. Beyond Elkhart (shaped like an elk's heart) the road winds through a gravel pit before entering the lodgepole pine and turning right near Bobs Lake.

Near the bottom of the first short hill is a crossroads (km 5.7). The road to the south (right) leads a short distance to Bobs Lake. The road straight ahead continues to Paradise and Island lakes, while the road(?) to the left (north) winds along a glacier-carved ridge to Boot Lake.

Calling the route to Boot Lake a " road" is stretching the meaning of the word. Unfortunately, the first half kilometre is deceptive. Not until you start sliding down two muddy ruts toward a bog do you realize that you should have turned around back at that last clearing between the scarred trees. Winter is another story. With a metre of snow on the ground, snowmobilers enjoy the maze of roads that link the lakes. (We have spent several enjoyable days exploring the area on touring skis.)

Prior to the Okanagan Connector, many of the roads in the area were little more than a series of loosely connected mudholes. Progress has changed all that — at least as far as the Paradise Lake Resort driveway at km 7.4. It is less than half a kilometre north to the public access to Paradise Lake or the recreation site at Island Lake, but the return to the "good old days" is noticeable in even that short distance.

It is worth the effort to dodge the last few mudholes as the Forest Service recreation site at Island Lake could be an excellent base to explore the rest of the upland lakes. Fishing is good in 46-hectare Island Lake, says one source, with rainbow trout to one half kilogram.

However, that wasn't always the case. "There were no fish in Boot Lake, Island Lake, Bobs Lake or the small lakes between Bobs Lake and Elkhart Lake," says Mountford. "Earl Martin was our game warden in the early years. He ran the eyeing station at Pennask Lake where they collected the rainbow trout eggs for the hatchery program. Earl used to keep a few eggs and hatch them, but he had no facilities to feed the fish. As soon as they were free-swimming fry, I loaded them up in cans and packed them on my back to stock these lakes. There was plenty of feed in Island Lake and Boot Lake and in two years they grew to eight pounds. We had some really fantastic fishing there for a few years."

While Gordon Mountford has retired to Creston, he did leave a legacy in the lake names on the upland plateau — one that may be difficult to pin on him in future years. It seems that Johns Lake, to the southeast of Paradise Lake, is named after one of the fishing camp's customers. "He was John Weithman from the Wenatchee area," says Mountford. "He owned a big orchard there and would come up five or six times a summer." And there is another lake to the east of Johns Lake that is named just that, "Another Lake." The story here is that young Jim Mountford was exploring the area when he discovered a new lake hidden among the pines. He raced back to the fishing camp to announce that he had found "another" lake — and the name stuck.

Figure 24
Paradise Lake, near the Elkhart Interchange. (Photo NVG-1084)

Elkhart Interchange also provides access to Elkhart Lodge, complete with a restaurant and gasoline station. If you have a good map and a little bit of luck, you can follow a series of old (and new) roads southeast to the Headwaters Lakes area.

The Sunset Interchange, about 12 kilometres east of Elkhart, offers access to Sunset Lake as well as Reservoir, Walker and Skunk lakes. Sunset is a walk-in lake and the others may as well be. There is a very rough road into the north end of 73.5 hectare Reservoir Lake, but for most fishers it could be a lot faster to walk the last 20 to 25 minutes to the lake.

In summary, with the elevation of the plateau ranging from 1500 to 1740 metres, the summer season can be short and unpredictable. Expect snow any time of the year, bugs in the spring and good fishing throughout the open season. However, if this is your idea of a vacation, you have found " Paradise."

•••

24

Princeton to Merritt

Highway 5A and 5A/97C

Statistics	**For maps, see pages 4, 85, and 110.**
Distance:	89 kilometres, Princeton to Merritt.
Travel Time:	One hour.
Conditions:	Paved throughout.
Topo Maps:	Princeton, B.C. 92 H/SE (1:100,000). Tulameen, B.C. 92 H/NE (1:100,000). Merritt, B.C. 92 I/SE (1:100,000).
Forest Maps:	Merritt Forest District Recreation Map.
Communities:	Princeton, Aspen Grove and Merritt.

Princeton lies 289 kilometres east of Vancouver, at the junction of Highways 3 and 5A. Princeton, originally called Vermilion Forks, has long been a mining community. The Natives mined the nearby red ochre deposits to make paint for trade with the coastal tribes. With the gold rush of the 1860s, Princeton became the gateway to the Boundary District. Coal was soon discovered beneath the streets of the town; then copper became the source of revenue. Similko Mines, south of Princeton, has worked a portion of the rich copper belt that extends from Chelan, Washington, to Cache Creek, B.C.

At the outskirts of lower Princeton, the inscription on a large granite block reads:

MILE ZERO No. 5 HIGHWAY
AUGUST 22, 1967.
PRINCETON, B.C.

The completion of the Coquihalla Highway in 1986 pre-empted the "Highway 5" name and the route was renamed "Highway 5A." A short distance north of the Mile Zero marker, Highway 5A crosses the Tulameen River then passes a junction to the left. Tulameen Road heads northwest, through Coalmont and Tulameen to Otter Lake. Otter Lake Provincial Park, a 45 unit campground, is located 33 kilometres from Princeton. See *Otter Valley Road*, page 113, for details.

Figure 25
Arrowleaf Balsamroot blossom with a bumblebee. (Photo WAB-144)

From Princeton north to the Missezula Lake junction, Highway 5A winds across rolling benchland covered with bunchgrass, stands of lodgepole pine and the occasional long-needled ponderosa pine. The glacial deposits are a mass of yellow each spring with the sunflower-like blossoms of arrowleaf balsamroot. To the east is Sagebrush Downs, the community racetrack and fair facilities. Farther east, you will see the line of the former Kettle Valley Railway (KVR) as it winds up the hillside on its way to Summerland.

A junction nine kilometres north of Princeton marks the start of a gravel road that leads 32 kilometres north to long, narrow Missezula Lake. Also at this point, Highway 5 leaves the grasslands and begins a steady climb through the narrow Allison Valley. It briefly opens into hay fields and picturesque Allison Creek, complete with beaver dams and yellow water lilies.

The Allison Creek fish barrier at km 16.9 marks the start of a chain of small lakes and reed-covered ponds that extend most of the way to Merritt. Fishermen and ducks are plentiful on these lakes. Deer are also so plentiful as to be a road hazard in the evenings. It's not uncommon to see the eyes of a dozen or more deer peering at you as you drive by after dusk. It is also not uncommon to have to brake to avoid hitting them as they wander across the highway.

On one rather cold, damp November day, I had the experience of a lifetime near here. I was travelling south from Merritt to Keremeos when I came over the crest of a hill and saw a gray lump on the road in my lane. I stood on the brakes and slid to a halt a few metres from the object. It turned out to be a very large, unconscious loon. Loons are helpless out of water, and this one must have assumed that the icy wet road was water. It had probably been knocked out on landing — now it lay there, not moving.

I took my old jacket and carefully wrapped the big bird, placing it in a cardboard box on the back seat of the car. Destination: open water. There wasn't any before Princeton, so we decided to head for the pools at Bromley Rock on the Similkameen River.

Unfortunately, the loon wasn't keen on waiting. The warmth of the jacket and the car revived the bird and it made a noisy attempt to get out of the box. The next half hour varied between comic and panic as we tried to keep the powerful bill from creating havoc while still keeping our vehicle on the busy highway.

At Bromley Rock, I carefully unwrapped the fully revived loon and gently lowered it into the water. It skittered two metres out on the pool, stopped and turned to look at us, called its haunting call, then bobbed downstream as though nothing had happened.

Allison Lake Provincial Park, at km 28.6, is divided by Highway 5A. Above the road is the campground, while below the road is a boat launch, picnic tables and beach. Many of the camping units of this small park overlook the lake with a thicket of Douglas fir absorbing the traffic noise of the nearby highway.

In the mid-1960s, my family and I spent our vacation at Sky Blue Lodge at the north end of Allison Lake. The lodge and cabins have long since disappeared and are being replaced by a cottage subdivision. A few minutes north of Allison Lake a two-kilometre-long hill climbs to the plateau country.

Hornet Lakes Road, a gravel forestry road to the right near the 35K signpost, climbs steeply up Missezula Mountain. This is the beginning of a network of old logging roads that are best explored with a four-wheel-drive vehicle, topographic maps, a good compass and lots of time. With care, you can follow the maze north past a series of potholes to Highway 5A at km 48. Watch for logging trucks.

Figure 26
A fly fisherman on McCaffrey Lake, north of Princeton. (Photo FIS-017)

Another junction, at km 37, marks a backroad to the AP Ranch resort and the Gulliford Lakes area. The Forest Service maintains a dozen recreation sites on a group of lakes lying west of the highway. A gravel road links Highway 5A with the Otter Valley. (See the *Pike Mountain Area*, page 110.) In addition to the fishing and camping opportunities of the area, there are also some challenging backroads. The AP Ranch junction also marks the southern border of the Nicola Valley cattle country. To the north, the vista is aspen and pine dotted hillsides, bunchgrass, cattle, ducks and the occasional moose.

Duck-covered ponds on either side of the highway and a cattle guard are all that mark the junction of Dillard Creek Road at km 48. Missezula Lake is 10 km southeast and fishing, canoeing and cool-water swimming await you. Watch for logging trucks and check the ruts and boulders before taking a heavy trailer down to the Forest Service recreation site at the north end of the lake.

Otter Lake Park is 43 kilometres south of the Highway 5A and Coalmont Road junction at km 54.5. Coalmont Road, also known as the *Otter Valley Road* (see page 113), rejoins Highway 5A in Princeton. Side trips along it include the historic Granite Creek gold and platinum placer diggings near Coalmont, and a backroad that runs from downtown Tulameen over the mountains to Coquihalla Highway 5 near the toll booths. (See *Coquihalla-Tulameen Road*, page 41.)

Bates Road, one kilometre farther north, leads to Kentucky-Alleyne Provincial Park. A score of fine upland lakes lie at the end of the six kilometre drive to the northeast. Two round ponds separate Kentucky and Alleyne lakes. In high water, these can provide swimming as well as canoeing and fishing. BC Parks has a 61-unit campground with a good walking trail around Kentucky Lake (see page 123).

A major junction at km 62.8 marks the Okanagan Connector (Hwy 97C). Phase III of the Coquihalla Highway system, this 90-km-long four-lane divided highway shortened the drive from Merritt to Kelowna by three hours. It had to climb to do it, though. At the summit of the highway it reached an elevation of 1728 metres, before making a 32 kilometre run down to Westbank.

Corbett Lake Country Inn's Peter McVey (km 70) has developed an reputation as a gourmet cook and expert fly fisherman. McVey trained in the kitchens of the Lord Mayor of London and worked in a number of large hotels before settling down to the quieter life at Corbett Lake.

Local skiers, in cooperation with the Forest Service, have developed 45 kilometres of cross-country ski trails in the Kane Valley area west of Corbett Lake. The trails are well marked with ample parking. Maps are posted at the parking lots and check-in stations. (See *Kane Valley Road* and *Kane Valley X-C Ski Area*, pages 83 and 89.)

Iron Mountain Road, at km 73, winds up into private ranchland on the eastern slopes of Iron Mountain. Logan's Marsh, across Highway 5A, is a wetland reclamation project undertaken jointly by Ducks Unlimited Canada and the Quilchena Cattle Company.

Lundbom Lake, four kilometres east of Highway 5A at km 77, lies in the upland grassland with open range on three sides and forest cover on the fourth. There are Forest Service recreation sites at the lake and plenty of old roads to explore. See page 128 for a map.

A beaver pond a short distance north of the Lundbom Lake junction marks the crest of the long descent to Coquihalla Highway 5 and Merritt. This was a 10 per cent grade before the road was reconstructed. If you are driving or pulling a heavy rig, this is still a good place to check your hitch and your brakes. The road descends almost 480 metres in the next seven kilometres.

A junction to the left just past the pond (km 80) provides access to the Old Nicola Road, a narrow 10-kilometre-long sidehill route to the foot of Nicola Lake. Access is from old Highway 5A opposite the brake check. Follow the old road down the hill for one kilometre and watch for a sideroad to the right just before a runaway lane. There used to be a sign marking this road, but it was missing in late 1997. If you reach the old viewpoint, you've gone too far.

Old Nicola Road passes under the main highway and then winds down past Hamilton Creek. This is no road for big-rigs or the weak-hearted traveller. This was the original 1870s wagon road from the village of Nicola to the mines at Granite Creek and it hasn't seen much improvement since.

If you are interested in a hike, a junction at km 2.4 marks the start of a rough road to Hamilton Lake and Sugarloaf Mountain (see page 127).

Coquihalla Highway 5 Interchange (Exit 286), at the bottom of the hill, is marked with a traffic light. The red-roofed log building nearby is the area travel information centre — one of the busiest in the BC Interior. In addition to racks of brochures and maps, the info centre has forestry displays and a small gift shop where local books and souvenirs can be purchased. The Godey Creek Trail network behind the centre is well worth exploring.

The interchange will require a decision. Straight ahead takes you past the Coldwater Road junction and into Merritt. A right turn and you are on your way to Kamloops or Nicola Lake while a left turn takes you south on Coquihalla Highway 5 to Hope and points west.

If you've come up from the Coast and are looking for an alternate route back to Hope, you might consider taking the Coldwater Road as far south as the interchange south of Kingsvale (Exit 256). This route is paved most of the way, with a few interesting diversions including native artist "Opie" Oppenheim's studio at the Coldwater village. Opie's original silkscreen paintings have an unique style that has garnered him considerable attention.

The traffic lights at the corner of Nicola Avenue and Voght Street in downtown Merritt mark another important junction. Straight ahead and Highway 8 will take you to Spences Bridge and the Trans-Canada Highway. Turn to the right and you will continue on Highway 5A past another chain of lakes to Kamloops and the Trans-Canada Highway. Why not stay a day or two while you decide on your next destination?

•••

25

Pike Mountain Area

Statistics	**For maps, see pages 4 and 110.**
Distance:	16.7 kilometres, Highway 5A to Coalmont Road.
Travel Time:	One to two hours.
Condition:	Gravel road.
Season:	Best in dry weather.
Topo Maps:	Aspen Grove 92 H/15 (1:50,000). Tulameen 92 H/10 (1:50,000).
Forest Maps:	Merritt Forest District Recreation Map.
Communities:	Merritt, Princeton and Tulameen.

The upland plateau in British Columbia's south Interior has many get-away places for back-country explorers and those who practice the fine art of fly-fishing. One of these is the Pike Mountain area, approximately half-way between Princeton and Aspen Grove. While Mount Pike is the focal point of the area, it is not the mountain that attracts outdoor recreationists. Instead, it is the more than two dozen lakes that lie hidden in the upland timber either side of Highway 5A.

The string of lakes along Highway 5A between Princeton and Aspen Grove hint of this hidden wealth. However, it is not until you venture off the highway that you discover the true recreational resource that lies in the folds of the plateau.

The AP Ranch sign at the junction of Pike Mountain Forest Service Road and Highway 5A, approximately 38 kilometres north of Princeton, is a good km 0 reference point. The junction to Andre Patry's AP Ranch is less than a kilometre west of the highway.

Every family unit has a member that is not enthralled by watching hand-tied flies sinking slowly into the wet. For these non-believers and others less inclined toward roughing it, the Patry family has established the AP Ranch on a homestead first settled in 1911. The main lodge has six rooms with expansion underway at the time of writing. They also have cabins that can sleep up to five persons. In winter they offer cross-country skiing, tobogganing, and ice skating on the nearby lakes. Summer activities include horse riding, hay rides, mountain bicycling and fishing. Contact the Patry's at (250) 378-6520 for more information.

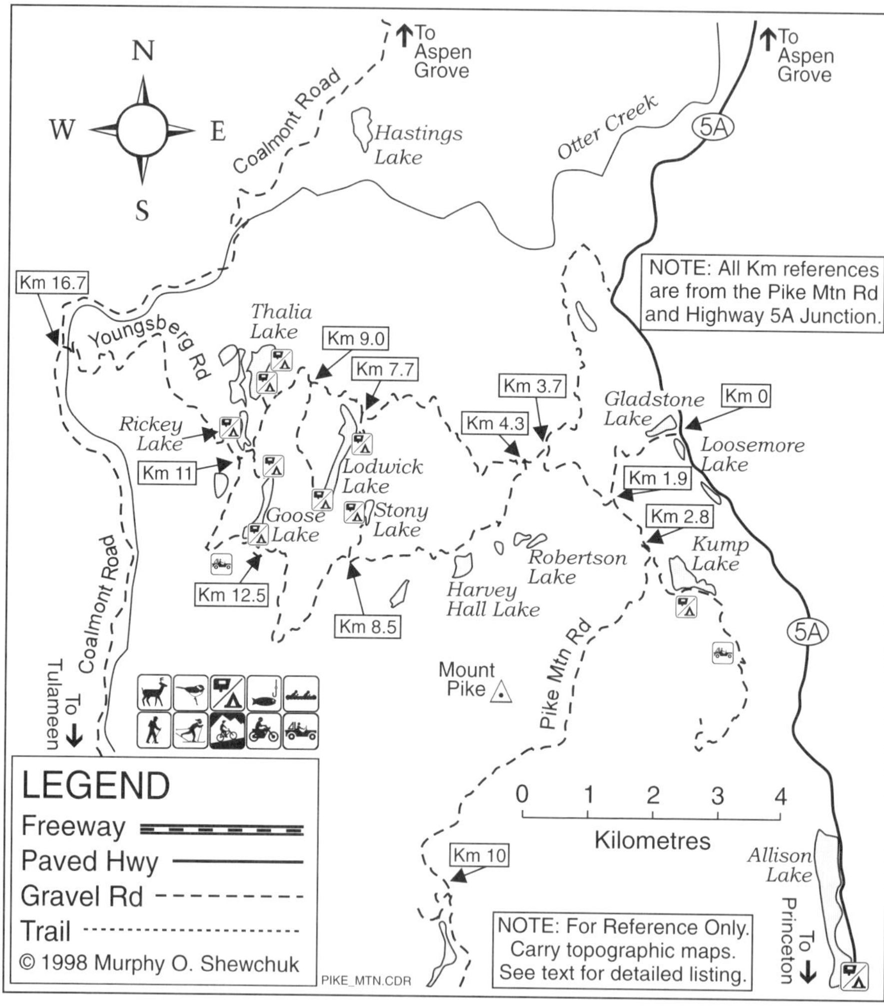

Map 20
The Pike Mountain area — Highway 5A to Coalmont Road.

The south (left) fork of a Y junction at km 1.9 presents an option for some serious back-country exploring away from the crowds (except for loggers and logging trucks). The first major diversion is again to the left at km 2.8. If you climb up the road to left, keeping right at the first junction, you'll soon find yourself following a narrow road around the south shore of Kump Lake. There is a small recreation site on the lake and a couple of other informal locations to park for a picnic or a little fishing. If you continue past the lake, the road gets progressively narrower until it ends in an old log landing near the summit of a ridge overlooking the Allison Creek valley.

When you are ready for more exploring, you can follow Pike Mountain road at least another 30 kilometres south before ending in a log landing in the lodgepole pine forest. There are a few small lakes along the way and multitude of side roads — take along a map and compass.

Figure 27
Goose Lake South FS Recreation site. (Photo PRN-359)

Meanwhile, back at the Y junction at km 1.9. This time, follow the road as it swings west and then south. A nearby sign marks this as the Robertson Lake Road, but this strip of gravel suffers from a bit of an identity crisis, for at a junction 1.8 kilometres later there is a "Pike Mountain Road" sign, a hold-over from when the road wound through the Pike Mountain Ranch instead of joining Highway 5A near Gulliford Lake.

The road the west at this junction climbs steadily up into the plateau, briefly breaking through the timber into open grasslands near the crest. A third junction part way up the hill marks the start of a road that loops south past Harvey Hall Lake and then north to the foot of Goose Lake before returning to the main route at Rickey Lake. (See the map on page

110 for details.) The route to Goose Lake is at best a rough two-wheel-drive road. From Goose Lake to Rickey Lake, there are some very narrow, steep sections that are certainly not conducive to anything but a four-wheel-drive vehicle.

If you are looking for a less strenuous route to the plateau lakes, continue west through the grassland meadows to the summit and then watch for a glimpse of the north end of Lodwick Lake near some ranch buildings. There is a recreation site among the timber on the northeast shore of Lodwick Lake and another, more open, site at the southwest end of the lake. Access to both is quite reasonable in dry weather.

Lodwick (or Ludwig) Lake is typical of many in the area. At 19.4 hectares, it is one of the larger, but like most, it is fairly narrow, lying in a north-south trench carved by glaciers. At an elevation of 1125 metres, it is a productive fishery and is stocked with rainbow trout. The east shore is primarily a rock cliff, but the west shore rolls into the upland mixed forest. In addition to fishing, it lends itself to canoeing and birdwatching.

The road continues west beyond Lodwick Lake before looping south around Thalia Lake. Just before Rickey (or Rick) Lake, a short side road extends south to a small recreation site at the north end of Goose Lake. Access to the open water is through a marsh and this lake may be a better bird-watching site than fishing hole.

Rickey Lake is the last in the series along this stretch of backroad. After a short climb through the timber, the road begins a long, steep descent to the Otter Valley, reaching Coalmont Road at Youngsberg Road, approximately 27 kilometres north of Tulameen.

•••

NEW

26

Otter Valley Road

Statistics	For maps, see pages 85, 110 and 118.
Distance:	80 km, Princeton to Hwy 5A near Aspen Grove.
Travel Time:	One to two hours.
Condition:	Partly paved, some rough gravel sections.
Season:	All season. Beware of mud in spring.
Topo Maps:	Princeton 92 H/SE (1:100,000). Tulameen 92 H/NE (1:100,000).
Forest Maps:	Merritt Forest District Recreation Map.
Communities:	Princeton, Coalmont, Tulameen & Merritt.

Ghost towns, a lost bucket of platinum, fighting trout and a quiet lake-shore campground are only a few of the attractions along the Otter Valley Road. For many years this 80 kilometre long byway was the main thoroughfare linking the mining centres of the Similkameen district with the ranching settlements of the Nicola Valley. Packers used the road to carry gold and platinum from the Granite Creek mines to the Canadian Pacific Railway at Spences Bridge, returning with supplies for the settlers and mining camps. The completion, in 1916, of the Great Northern and Kettle Valley railways took the freight off the Otter Valley Road. It was relegated to backroad status when the present route of Highway 5A via Allison Lake was completed in the 1920s.

The earliest records of the passage of white men through this valley date back more than 185 years. Alexander Ross of John Jacob Astor's Pacific Fur Company is believed to have followed Indian trails through here in January 1813. Ross and his companion were returning to the Columbia River after having just established a post at the junction of the North and South Thompson Rivers — now Kamloops.

In 1826, Archibald McDonald, a Hudson's Bay Company clerk from Fort Kamloops, travelled through the area and recorded the lay of the land in "A Sketch of Thompson River District-1827". On his map, McDonald named what we now know as the Similkameen River the "Schimilicameach", and the Tulameen River the "Red Water Branch".

The red reference again turned up in the early name for Princeton — Vermilion Forks — which referred to a deposit of red ochre in the bluffs a short distance up the Tulameen River. The Forks soon became a strategic point in east-west travel following the discovery of gold in the Boundary Region in October of 1859. Princeton received its present name when, in 1860, Governor James Douglas renamed it in honor of the Prince of Wales — later Edward VII — who visited Canada that year.

Today the traveller can gain access to the Otter Valley Road by leaving Highway 5A in Princeton and crossing the old wooden bridge at the end of Bridge Street. Follow the signs marked Tulameen Road. It may seem a bit confusing but at the north end, where this byway rejoins Highway 5A, signs mark it the Coalmont Road. The original settlers knew it as the Otter Valley Road, a name worth sticking with rather than trying to settle an argument.

The road leaves Princeton and follows a scenic twisting route high above the Tulameen River canyon for nearly 19 kilometres before gradually dropping back down to river level at Coalmont. According to Mollie Broderick, whose grandparents were early settlers in the Otter Valley, Bobby Stevenson, of Barkerville and Granite Creek fame, was the surveyor of this portion of the road and "he just followed a garter snake along the hillside to find the grade".

Coalmont, for many years a thriving railway "port" for the Blakeburn coal mines, is now trying to build a base in today's diverse economy. Blakeburn, on the hillside 9.6 kilometres from the Coalmont Hotel, was the source of Coalmont's wealth, its coal rated as "amongst the finest in the continent." See *Coalmont to Lodestone Lake*, page 117, for more information on Coalmont, Blakeburn and nearby "Granite City".

On the 6.5 kilometres from Coalmont to Tulameen, Otter Valley road again climbs well above the river, offering an excellent view of the bottomland with the former KVR right-of-way winding through it like an endless snake.

Tulameen, first known as Campement des Femmes, then later as Otter Flats, was important for the commerce of the Interior between 1849 and 1860. As part of the Hudson's Bay Company Brigade Trail system, it received extensive use until the completion of the Dewdney Trail and the Cariboo Wagon Road provided easier alternate routes.

These fur brigades were no small venture. Early records show that as many as 400 horses and 50 men made up the convoy. Each of the pack horses was laden with two 35 kilogram bales of fur. On the return trip they carried supplies and trade goods.

Tulameen may be a good place to stock up on fuel and supplies before exploring the high country to the southwest or the backroad to Coquihalla Lakes. On the northern outskirts of Tulameen a public boat launch provides access to Otter Lake. Part way up the west side of the lake is Otter Lake Provincial Park, on a tree-covered point.

Forty-two kilometres north of Princeton and 15 kilometres north of Tulameen, Otter Valley Road passes the gate of Thynne Lake Ranch. No story of the road would be complete without a mention of Jack and Mary Thynne (pronounced "thin"). John Granville (Jack) Thynne came from an aristocratic background, a direct descendant of Sir Richard Grenville. His wife Mary Eliza was born the third daughter of William Linklater, a free-trader, at Fort Garry — now Winnipeg, Manitoba. They were married at Pelly, Saskatchewan in 1887. After an attempt at homesteading on the prairies and a ranching partnership in the Nicola Valley, they settled in the Otter Valley in 1890.

At that time there was only a trail from Nicola to Granite Creek so Mary carried her two-year-old daughter Ethel with her on her saddle horse for the 70 kilometre journey. Soon after their arrival, Jack Thynne circulated a petition to have a wagon road built from Nicola to Granite Creek. With the Canadian Pacific Railway at Spences Bridge and a road from Nicola to Spences Bridge, this appeared to be a logical route for the commerce of the Similkameen district and the petition met with success.

To serve travellers on the new road, the Thynnes erected a two-storey building that served as home and "Thynne's Road House". Travellers were treated to "a little bit of old England" when they gathered in the sitting room to enjoy conversation amid portraits and photographs of the old Thynne home in Cornwall, England. At the opposite end of the room stood a grand piano which daughter Ethel played as the guests gathered around to sing, accompanied by Jack Thynne on his violin or banjo.

Until the arrival of the Vancouver, Victoria and Eastern Railway (VV&E) in 1915, the Thynne house served the needs of travellers.

Among those who visited this well-known Otter Valley stopping place was a slender man, obviously in his mid-50s, who spoke with a southern accent and enjoyed music and dancing. His name, he said, was George W. Edwards. Mr. Edwards lived on a ranch eight kilometres out of Princeton and made frequent trips which few people cared to ask him about. On one of those trips he was unmasked while holding up the Canadian Pacific Railway at Ducks, about 25 kilometres east of Kamloops. Edwards and his partners were caught a few days later on May 14, 1906, in the Douglas Lake country. Residents of the Nicola and Similkameen country were surprised to learn that their friend was none other than the notorious Bill Miner, train robber. Other well-known, but not so notorious guests included Pauline Johnson, Father Pat, and Walter Camsel.

For the next 22 kilometres the road winds its way up Otter Creek and out on to the grasslands plateau that is a trademark of the Nicola country. About 12 kilometres south of the junction of the Otter Valley Road with Highway 5A, it passes a side road that leads 16 kilometres northwest to Brookmere.

Less than one kilometre farther north, another side road leads to the Voght Valley and a series of popular fishing lakes equipped with Forest Service recreation sites. As the road continues northward, the reason why ranching is an important pursuit in the Nicola Valley becomes increasingly apparent. The lightly-timbered parkland gives way to natural open grassland. Bunchgrass-covered slopes, bright yellow with the blossoms of arrowleaf balsamroot in early spring, green in early summer and golden yellow by autumn, provide the basis for an industry that ships 15,000 head of cattle each year.

Scattered lakes and ponds provide natural reservoirs for life-sustaining water that serves wildlife as well as the cattle industry. Many of the larger lakes have been stocked with rainbow (Kamloops) trout and are popular spots for sports fishermen.

Eighty-kilometre-long Otter Valley Road, called "Coalmont Road" at the north end, follows a route first established by nomadic Indians. It has been gradually improved by fur traders, miners and settlers until today one-third of it is paved and the remainder is gravel. If time is of little importance, Otter Valley Road offers a glimpse of our varied and colourful past. If speed is the main consideration, Highway 5A via Allison Lake is the way to go.

•••

NEW

27

Coalmont to Lodestone Lake

Statistics	For map, see page 118.
Distance:	23 km, Coalmont to Lodestone Lake.
Travel Time:	One to two hours under good conditions.
Elevation Gain:	1080 metres.
Condition:	Gravel road, some very rough and narrow sections.
Season:	Mid-July to mid-October.
Topo Maps:	Tulameen, B.C. 92 H/10 (1:50,000) Princeton, B.C. 92 H/7 (1:50,000)
Forest Maps:	Merritt Forest District Recreation Map.
Communities:	Coalmont and Tulameen.

My first trip from Coalmont to Lodestone Lake, 25 kilometres due west of Princeton, was a one-way affair. That's not to say that I never came back, but that I used a more strenuous and round-about way to do it. This was on a Saturday morning in late July, 1980. And I was under the leadership of Randy Manuel and the late Victor Wilson, both of the Penticton area. Unfortunately, my view from the back seat of the 4x4 crew-cab was little more than the back of someone's head. My experience, instead, was based on feeling — a climb straight up a mountainside. After our two-hour drive from Princeton, we took a walk across the Cascade Mountains on the Hudson's Bay Company Brigade Trail. When we reached Peers Creek the following Saturday, I was prepared to gladly accept any seat in any vehicle.

My most recent trip was much more comfortable as I was in the driver's seat of my 4x4 VW van. Well-used logging roads make up much of the route from Coalmont to Lodestone Lake. However, from km 13 to km 18, the brush is reclaiming the road in several places and a sharp machete may be necessary to protect you paint job. From km 18 to km 21, it still feels like a climb straight up a mountainside. Only this time I could see why it felt that way — and it wasn't my imagination.

Consider this preamble a subtle hint. A four-wheel-drive vehicle with good clearance is essential. The afore-mentioned machete, an axe (or chainsaw), a shovel, plenty of fuel and a little bit of foolhardiness are also good tools for this trip.

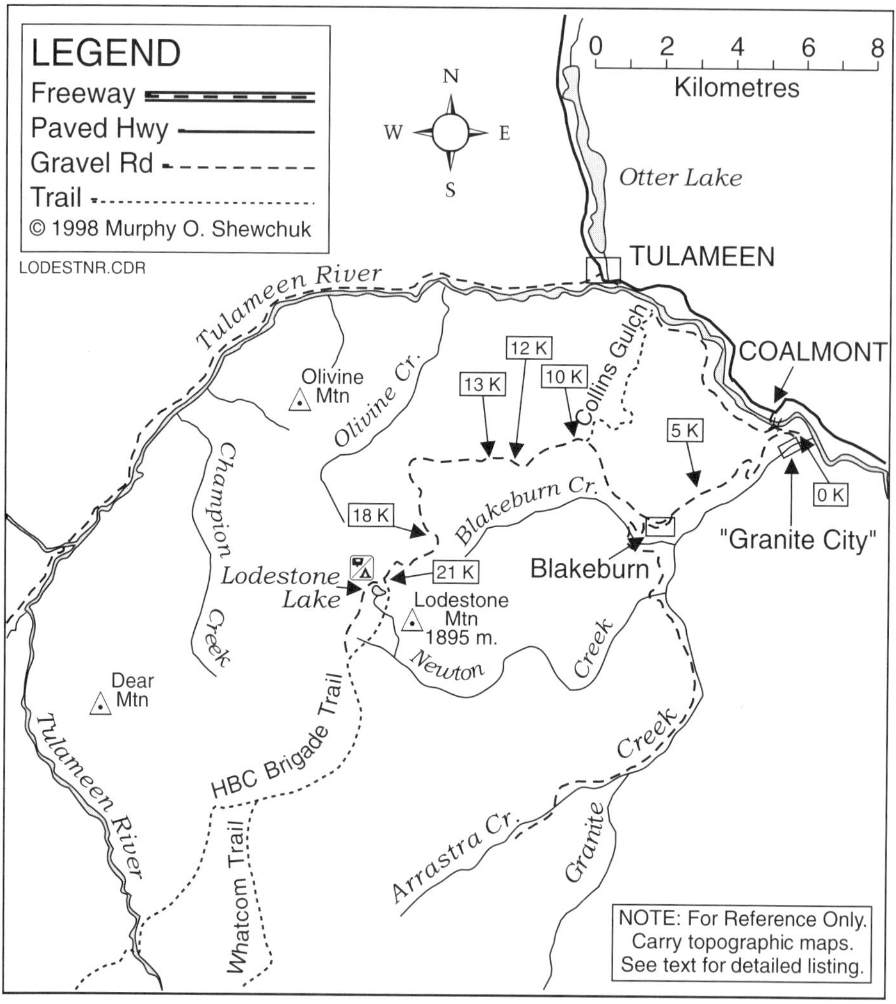

Map 21
Tulameen-Coalmont-Lodestone Lake area.

The starting point for this description is the 4-way stop at the Coalmont Hotel, on the highway between Princeton and Tulameen. Coalmont Collieries originally established the townsite in 1911 as a base for a coal mine at Fraser Gulch, to the west across the Tulameen River. Although the deposit was substantial — they found seams up to 5 metres thick, 600 metres into the mountain — the coal was too crumbly to be mined profitably. Interest shifted to the coal deposits on the north fork of Granite Creek (see Blakeburn) and the settlement that was once expected to have a population of 10,000 barely survived.

The Coalmont Hotel, constructed in 1911-12, was the community centrepiece. This handsome three-storey woodframe structure is being brought back to its former glory by Donna and Peter Jack. The Jacks took over the premises in September 1995, and have been restoring the building from a new foundation up. As renovations proceed, the completed rooms are serving visitors from around the world. When Peter Jack completes the refurbishing job, he expects to have 24 hotel rooms true to the pre-World War I style.

The first leg of this backroad tour is a short one. Go south (towards the Tulameen River) for about 0.3 kilometres, then east for about 0.2 kilometres to the Bailey bridge across the Tulameen. After crossing the river, go left (southeast) for 0.8 kilometres to a sharp right turn. Check for logging trucks, and when it's safe, angle across the road to the open area with the decaying log cabins.

A BC Forest Service recreational site marks the remnants of the community of Granite Creek. In 1886, promoters boasted that it was the largest British Columbia settlement outside Vancouver, Victoria and New Westminster.

Historians suggest that John Chance, a sometime cowhand, miner, cook and rustler, discovered a placer gold deposit on Granite Creek on July 5, 1885. By that October, 62 companies were working the diggings. Despite high water and a shortage of lumber, they took out $90,000 in gold that year. On December 18, 1885, mining recorder Henry Nicholson reported "Granite City is rapidly increasing, buildings of one kind or another number two hundred."

Nearby Hines, Slate (Olivine), Champion and Bear Creeks also produced favorable diggings. With the coarse gold valued at $17 per ounce, miners recovered a nugget worth $87.50 from Granite Creek while Bear Creek yielded a gold nugget worth $320.

Intermixed with the gold was a heavy whitish metal that was difficult to separate from the gold. According to N.L. "Bill" Barlee, when a miner named Johanssen left his claim, he cached a pail containing 300 ounces of the "worthless" stuff somewhere near Granite Creek. A short time later, assayers in England discovered that this nuisance metal was platinum — today worth more than gold. By 1891, the district had become the most important platinum producer in North America.

Incidentally, Bill Barlee suggests that Johanssen never returned to dig up that the bucket of platinum.

Five years after the initial discovery, the gold-rush had slowed to a walk. An 1894 report indicates that only a few large companies were still working the gravel of Granite Creek.

Today, weeds, wild roses and saskatoons hide most of the streets of Granite Creek. A concrete monument, a bumpy dirt road and a few decaying buildings are all that remain to mark the creek-side community that yielded more than $2 million in gold and platinum.

Figure 28
Mine diggings at Granite Creek. (Photo PRN-406)

After you've had an opportunity to look around, head back out to the junction at the corner and this time continue up the hill on what Tolko Industries has labelled "Blakeburn Road". For the sake of simplicity and this description, this would be a good place to reset your odometer. You should pass some beautifully landscaped buildings at km 2.7 and reach a junction with Arrastra Creek Road near km 6. Arrastra Creek Road can take you a considerable distance south into the mountains, but if snooping around old mines is your pleasure, you could get your fill in less than a kilometre.

According to Don Blake, in *Blakeburn: From Dust to Dust*, Henry Lowe discovered what was to become the "Blakeburn" coal deposit in 1906. Coalmont Collieries, formed to develop coal deposits in Fraser Gulch, began work at the mine site in 1913, but quit during the First World War. By 1915, the Geological Survey of Canada reported an estimated 65 million tons of commercial quality coal in the Tulameen coal basin. W. J. Blake Wilson and Senator Pat Burns (of meat-packing fame) bought Coalmont Collieries in 1917 and secured a lucrative contract with the newly built Kettle Valley Railway (KVR).

In 1918, foolhardy drivers moved over 5000 tons of coal down the mountainside in five-ton "auto-trucks" equipped with solid rubber tires-and poor brakes. They even tried horse-drawn sleighs in winter. It soon became clear that both methods were unsafe and inefficient.

The company completed construction of an aerial tramway in 1920, linking the "Wilson Portal" with the rail yard at Coalmont. An incredible feat of engineering for the time, a bucket-laden with a ton of coal-left the portal every 55 seconds for the Coalmont KVR siding 4.5 kilometres distance. The 250 buckets were specially built mine cars that rode on wheels and axles underground and hung from a frame as they rode the cable down to Coalmont. The weight of the full buckets descending about 390 metres in altitude carried the returning buckets up at a pace slightly faster than a person could walk. The returning buckets weren't always empty. The aerial tram carried up groceries and supplies for the company store and mine workshops.

In their peak year (1928) the 300 Blakeburn miners produced nearly 165,000 tons of coal, rated as "amongst the finest in the continent." The town boasted a population of more than 500 people. The miners and their families were proud of their school (with 100 students), library, tennis court and the baseball field cut out of the hillside.

An underground explosion at 6:30 p.m. on Wednesday, August 13, 1930, marked the beginning of the end for Blakeburn. Of the 46 men on shift, only one escaped alive. Subsequent investigations were unable to determine what triggered the explosion although all agreed that a violent thunderstorm passed through the area at the time.

Soon after "Black Wednesday" the coal market began to change and production declined until the last shift turned in their lamps at 4:00 p.m. on April 8th, 1940. By the end of the summer, the town was deserted.

Lodestone Road, to the right near km 6, climbs northwest, passing several open coal seams and more remnants of the Blakeburn diggings. Several little-used side roads near km 9.5 lead northeast to the head of Fraser Gulch and Collins Gulch. The Collins Gulch Trail was part of the original Hudson's Bay Company Brigade Trail from Otter Lake to Hope. It's about six kilometres down to the Tulameen River near Walt Smart's Ranch, but in mid-October 1996, the trail was in desperate need of re-marking and removal of hundreds of windfalls.

The next major junction on your route to the top is just past the 12K marker. To the left is Slate Creek Road. Keep right and continue west to the 13K marker. Here a lesser used road to the left follows the base of a valley upstream. In the fall of 1996, the brush was growing back in for the first couple of kilometres along the creek, then the road opened up into several old cut blocks.

So far the climb has been steady, but not particularly trying. However, a junction near km 18 (keep right) marks the start of a serious ascent to the 1800 metre ridge and Lodestone Lake. This section could have suffered from the heavy snow and spring rains of the 1996-97 season. However, if it is passable the subalpine meadows along the ridge can be a naturalist's paradise.

The HBC packers also considered this a paradise and chose it as their first camp after leaving Otter Lake and the Tulameen River. A rough road was constructed from Blakeburn to Lodestone Lake and south to Wells Lake and the Whipsaw area in the early 1960s. Rough is still the key word. A series of mudholes starting about 1 kilometre south of Lodestone Lake make this route virtually impassable except during extremely dry weather.

For most summer travellers, the Forest Service recreation site at Lodestone Lake can be an excellent base from which to explore the sub-alpine meadows — on foot.

Additional Information:
Coalmont Hotel,
Box 3A, Compartment 11, Coalmont, BC V0X 1G0
Phone (250) 295-6066 Fax (250) 295-0086

•••

GPS References for major points of interest. See page 9 for details.

Wpt	Description	Latitude	Longitude
W01	Coalmont Hotel	N49° 30' 37.218"	W120° 41' 34.088"
W02	Granite Creek Rec. Site	N49° 30' 11.233"	W120° 40' 55.492"
W03	Arrastra Creek Rd (Blakeburn)	N49° 29' 05.351"	W120° 43' 54.814"
W04	Slate Creek Rd Junction	N49° 29' 48.494"	W120° 47' 33.734"
W05	13K Junction	N49° 29' 53.573"	W120° 48' 12.010"
W06	18K Junction	N49° 28' 41.791"	W120° 49' 19.042"
W07	Lodestone Lake Rec. Site	N49° 27' 58.165"	W120° 50' 51.526"

•••

28

Kentucky-Alleyne Park

Statistics	For map, see page 124.
Distance:	6 kilometres, Highway 5A to Kentucky-Alleyne Park. 8 kilometres to north end of Alleyne Lake.
Travel Time:	10 minutes.
Condition:	Paved to main campground.
Season:	Year around.
Topo Maps:	Aspen Grove 92 H/15 (1:50,000).
Forest Maps:	Merritt Forest District Recreation Map.
Communities:	Merritt, Aspen Grove & Princeton.

Katharine paddled at the bow of the yellow canoe. Madeleine sat in the centre leaning up against the carrying yoke. I sat at the stern, paddle in hand, as we drifted toward the reeds at the north end of Alleyne Lake, intent on capturing a marsh wren on film.

Katharine's parents, Madeleine and Wilf, had ventured inland from their retirement home in Victoria and were spending a weekend with us in the beautiful Nicola Valley. While Wilf prepared the steaks, Katharine and I assisted Madeleine into the canoe for a bit of exploring. We were enjoying the late afternoon sun, oblivious to growling stomachs and the aroma of supper cooking on the campfire.

As the canoe parted the reeds on a muskrat waterway, Katharine stopped, pointing to a nondescript mound on the water's edge. I strained to see what had caught her attention and gradually made out the near-perfect camouflage of a nest. I quietly put down the paddle and as I focused the telephoto lens I saw the red eye of a loon.

Katharine held the canoe steady and Madeleine remained motionless a metre or so from the big bird. The only sound was the camera shutter, but it was enough to disturb the nervous loon. As we paddled backwards from the nest, it clumsily pushed its muscular chest over the bed of reeds and slid deep into the water.

We had only moved back a few metres when it surfaced, almost within reach of the canoe, and began a dance guaranteed to frighten all

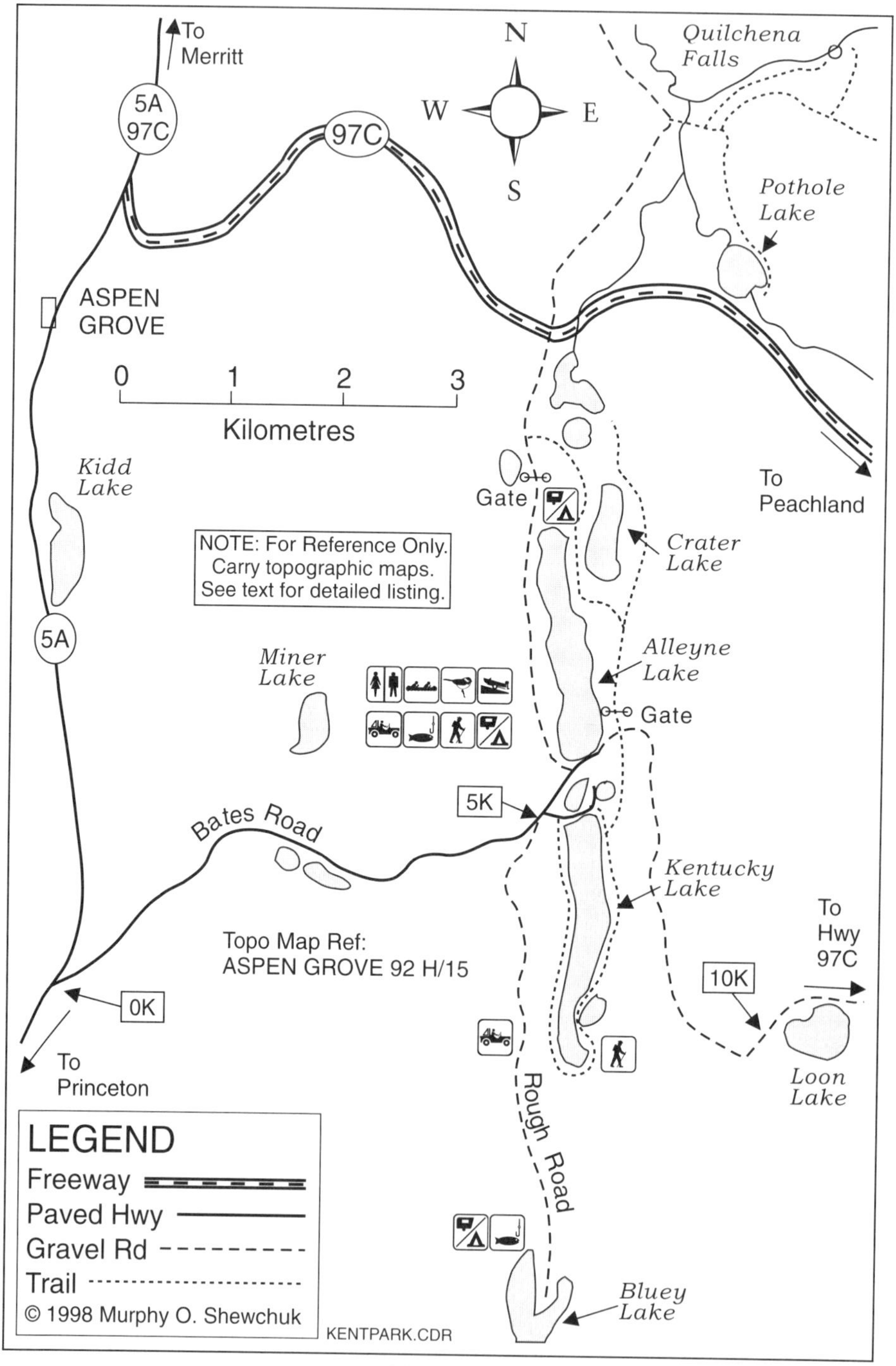

Map 22
Kentucky - Alleyne Provincial Park and area.

but the most determined of predators. While Katharine carefully guided us away from the loon's nest, I continued to photograph the splashing loon as it screamed and threatened us. Water flew everywhere. Twice it feigned an attack before retreating and again throwing a spray of water in our direction.

As I ran out of film, we slipped out of the loon's territory and the adrenaline levels in all four of us began to subside.

Two-kilometre-long Alleyne Lake lies south of Merritt and east of Aspen Grove, within sight of the Okanagan Connector. It is part of the Kentucky-Alleyne Provincial Park, established on March 5, 1981. The "Kentucky-Alleyne Park" signs at the Loon Lake Interchange on the Okanagan Connector or at Bates Road, on Highway 5A about 5 kilometres south of Aspen Grove, give no hint of treasures in the heart of the plateau. Alleyne and Kentucky lakes are the largest, with a busy provincial campground nestled among the trees between the two lakes.

Alleyne Lake, at 55 hectares, is the larger of the two main lakes while Kentucky Lake, at 44 hectares, follows a close second. Between the two lakes is a pair of ponds that are ideal training basins for kids and canoes.

Bluey Lake, a short drive south over a very rough road, covers an area of 28 hectares. Although not presently part of the park, there is also a Forest Service Recreation Site at the north end of Bluey Lake. The access road to the lake winds south from Bates Road, starting to the right just before the park entrance sign at km 5.0. The rough gravel road passes through an old log cut and alongside an ancient canyon, probably a major river at the end of the last ice age. At the time of writing there was a rough steep section at km 2.5 that would discourage all but the most ardent four-wheeler.

None of the lakes have natural spawning streams and instead must rely on the Fish and Wildlife Branch's stocking program. While fishers work Kentucky and Alleyne year around, pulling trout that run at 0.2 to 0.4 kilograms, loons also do their share of fishing. More than one fisher has pulled in a rainbow with telltale marks that indicate it barely escaped the grip of a deep-diving loon.

The surrounding land is a classic example of glaciation with the lakes and ponds set out in a north-south string. Alleyne Lake is in the transition zone between the dry grasslands and the slightly wetter pine and fir forest. Scattered ponderosa pine and interior Douglas fir cover the western slopes near the lake while grass dominates the northern and eastern slopes. Pine and fir surround Kentucky and Bluey lakes.

Several old roads encircle Alleyne Lake. North and east of the lake, they cross deeded land belonging to the Douglas Lake Cattle Company.

Fences and "No Trespassing" signs discourage the indiscriminate use of the delicate rangeland.

Figure 29
Canoeing on the ponds between Kentucky and Alleyne lakes.
(Photo KAA-180)

Kentucky Lake Trail

Wildflowers, wildlife, waterfowl, excellent fishing, one-metre-thick ponderosa pine and equally large Douglas fir are all features of a two-hour pocket-wilderness hike around Kentucky Lake. The well marked four kilometre loop trail begins near the campground on the northeast side of Kentucky Lake. Small red aluminum "flags" nailed to trees act as a guide in some areas, but the trail is virtually impossible to lose. It is an easy hike around the lake — pack your lunch, water bottle, camera and binoculars and turn it into a half-day saunter.

The completion of the Okanagan Connector (Highway 97C) has resulted in increased pressure on the resources of the Kentucky-Alleyne area. With care, common sense and government support, fishers, campers, hikers, ranchers and loons will continue to enjoy the rugged beauty, good fishing, excellent hiking and other recreational benefits of the upland lakes.

•••

NEW

29

Sugarloaf Mountain Ramble

Statistics	For map, see page 128.
Distance:	1.5 km, Hamilton Lake to Sugarloaf Mountain peak.
Elevation Gain:	Approximately 300 metres.
Travel Time:	Approximately 3 hours round trip.
Condition:	Indistinct trail. A few steep sections.
Season:	Summer and fall. Can be muddy in spring.
Topo Maps:	Merritt 92 I/2 (1:50,000).
Forest Maps:	Merritt Forest District Recreation Map.
Communities:	Merritt.

The hike up Sugarloaf Mountain from Hamilton Lake is an excellent afternoon ramble. It is particularly enjoyable in the early summer when the wildflowers are at their best. On one particular spring afternoon, we found 25 different species of wildflowers in blossom between the lake and the 1360-metre peak.

The sugarloaf shape of this mountain is clearly visible on the skyline to the east of Merritt. Access is from the Old Nicola Road which leaves the old Highway 5A just beyond the viewpoint at the top of Hamilton Hill. A rough road winds up Hamilton Creek from Old Nicola Road. Park near the mid-point of Hamilton Lake and follow the fence trail up to the eastern base of Sugarloaf Mountain. Keep to the right of the main rock bluff and follow a natural trail that traverses the rocky face starting near a pair of three-metre-high juniper trees.

The view from the top is excellent. In addition to the wildflowers, grouse, deer, bear and coyotes can sometimes be seen on the open rangeland. An alternate route down is via the "back" or north side of the mountain. You can work your way down through the trees to a grassland bench. Then follow the cattle trails around the west side of Sugarloaf back to Hamilton Lake.

Please do not take vehicles of any sort off the roads as the grassland is delicate and easily damaged.

•••

Figure 30
Hamilton Lake and the grasslands from Sugarloaf Mountain. (Photo NVG-902)

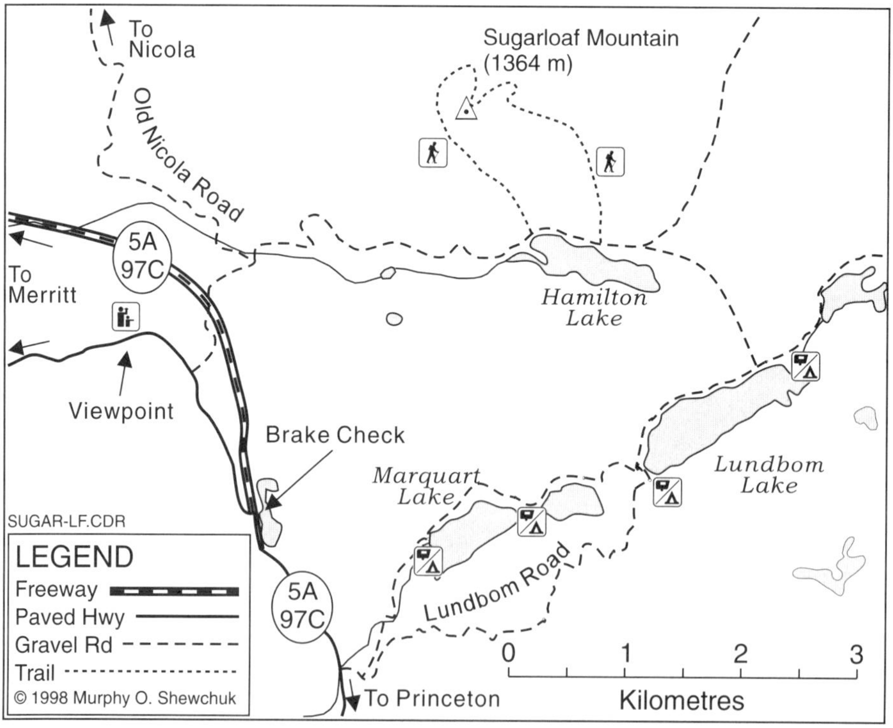

Map 23
Sugarloaf Mountain - Lundbom Lake area.

30

Merritt to Kamloops

Coquihalla Highway 5

Statistics	**For map, see pages 4 and 137.**
Distance:	80 kilometres.
Travel Time:	Less than one hour.
Condition:	Paved, divided highway.
Topo Maps:	Merritt 92 I/SE (1:100,000). Kamloops 92 I/NE (1:100,000).
Forest Maps:	Merritt Forest District Recreation Map. Kamloops Forest District Recreation Map.
Communities:	Merritt & Kamloops.

Exit 290, on the northeast side of Merritt, marks the junction to Highway 5A and Nicola Lake. It also signals the start of the long climb out of the Nicola Valley. A pull-out six kilometres north of the interchange provides a view of Nicola Lake and the pioneer settlement of Nicola. This view has changed little in the past century: cattle still graze on the grassland and the Nicola River still twists and turns along the valley floor.

Another 10 kilometres up the hill, the view disappears, blocked out by Mount Mabel as Highway 5 enters the confines of the Clapperton Creek valley. Helmer Road (Exit 315) is usually closed to traffic beyond the interchange loop in the summer months to protect the fragile upland environment. But in winter, the gates are opened to snowmobilers and 1723-metre-high Swakum Mountain becomes a beehive of droning machines.

Using the BC Hydro right-of-way, old roads and some especially cut trails, the Swakum Mountain Trail Riders Association and the Forest Service have created approximately 78 kilometres of interlocking trails on the mountain to the west of the highway.

Coquihalla Highway 5 crests at about 1450 metres elevation on the upland plateau about six kilometres north of Helmer Road exit. If you look carefully through the trees to the east, you may catch a glimpse of Surrey or Sussex lakes. These fine fishing lakes are not acccssible from Highway 5, but can be reached from the Logan Lake Road, west of Exit 336. (See *Surrey-Sussex Lake Road*, page 139.)

Although these huge animals don't usually stick around to be photographed, moose can often be seen in the marshes of the Meadow Creek Valley, between the crest and the Walloper Lake interchange (Exit 336). The chain link fences along the route in this area were installed to protect highway travellers and big game from each other.

Walloper Lake is just one of the reasons to leave the freeway at Exit 336. Lac Le Jeune Provincial Park and several Forest Service recreation sites also lie to the east of the highway, on or near the old road from Logan Lake to Kamloops. (See *Logan Lake Road,* page 135.) To the west is the access to Surrey Lake as well as the community of Logan Lake and the giant copper mines of the Highland Valley.

Also accessible from Exit 336, Lac Le Jeune resort offers accommodations, food services and a conference centre. The Stake Lake cross-country ski trails are also of major interest during the winter months.

The brake check pull-out marks the top of the long hill down to Kamloops and the Trans-Canada Highway (Exit 362). A word of caution as you approach the Trans-Canada Highway — the speed limit changes from 110 to 90 kph. As a second caution, watch the signs carefully as you continue northeastward. The Columbia Street interchange (Exit 369) can be confusing. Keep to the inside or centre lane if you want to bypass Kamloops and keep to the right lanes if downtown is your destination.

Exit 374 marks the departure from the freeway and the continuation of Highway 5 (The Yellowhead Route) toward Jasper. The four-lane highway continues eastward as the Trans-Canada Highway (1/97), becoming a two-lane highway east of Kamloops.

•••

31

Merritt to Ashcroft

Highway 97C

Statistics	For maps, see pages 4 & 137.
Distance:	44 kilometres, Highway 8 at Shulus to Logan Lake. 60 kilometres, Logan Lake to Ashcroft.
Travel Time:	2 hours.
Condition:	Paved throughout.
Season:	All seasons.
Topo Maps:	Merritt 92 I/2 (1:50,000). Mamit Lake 92 I/7 (1:50,000. Spences Bridge 92 I/6 (1:50,000). Ashcroft 92 I/11 (1:50,000).
Forest Maps:	Merritt Forest District Recreation Map.
Communities:	Merritt, Logan Lake & Ashcroft.

The Mamit Lake Road portion of Highway 97C, in addition to being the main link between Merritt and the copper mines of the Highland Valley, was once the route of gold seekers travelling from the Similkameen to the Cariboo.

Mamit Lake Road leads north from Highway 8 at the Indian community of Shulus, five kilometres west of Merritt. It passes through dry bunchgrass and sagebrush country before entering scattered timber as it climbs the Guichon Creek valley. Approximately four kilometres north of the Highway 8 junction, the hillside workings of the Craigmont copper mine can be seen across the valley. Craigmont, which closed in 1983, once employed over 600 workers.

Several side roads lead to ranches, lakes and logging operations on the plateau country to the east of Mamit Lake. The Mamit Lake area was the Nicola Valley's French district, first settled in 1873 by Joseph and Pierre Guichon and later by Louis Quenville and Jean and François Rey.

Joseph Guichon later established a ranch on the southeast shores of Nicola Lake and, in 1907-08, built the famous Quilchena Hotel.

Figure 31
Pottery on display at the Brookmere Pottery shop, located on Highway 97C near Mamit Lake, 23 kilometres north of Highway 8. (Photo NVG-1172)

Mamit Lake Road ends at the copper mining community of Logan Lake, where the Highland Valley Road, newly-designated as Highway 97C, continues to Ashcroft and other roads branch off to Savona and Kamloops.

Logan Lake is a community that has seen more than its fair share of ups and downs. Because of the mines, its fate has been tied to the vagaries of world copper prices. Logan Lake District Council has attempted to diversify the community employment base and the village tax rolls. Logan Lake has been successful in encouraging retired people to move into the relatively low-cost housing. It has also been successful in creating a recreational environment that would be the envy of other communities of similar size. (See *Logan Lake Road*, page 137 for additional information on outdoor recreation in the area.)

Cross-country skiing has become a major winter pastime. The cross-country ski trail network starts near a television tower above the village and extends to the north. At the time of writing, the Highland Valley Outdoor Association had 36 kilometres of groomed cross-country trails including two kilometres of lighted track. The system has something for everyone with beginner trails near the community and a mixture of intermediate and advanced trails throughout the network.

Highland Valley Road bisects North America's largest copper mine. With the first claims in the Guichon Batholith staked in 1896, mining and exploration proceeded sporadically for more than a half century before technological developments made mining of the low grade deposit profitable.

H.H. "Spud" Huestis and his associates formed the Bethlehem Copper Corporation in the mid-1950s. Bethlehem Copper began an intensive exploration program that established copper/molybdenum ore reserves exceeding 60 million tonnes at approximately five per cent copper. After several attempts at financing, Japan's Sumitomo Metal Company invested $5.5 million, and on 1 December 1962, Bethlehem Copper began production.

Egil Lorntzsen, another British Columbia mine-finder of note, first staked claims in 1958-59 on what became the Lornex Mining Corporation property (now Highland Valley Copper). It took more than a decade before a $144 million construction program led to Lornex's first shipment of copper concentrate in July 1972. In its first decade of operation, Lornex expanded to become a giant open-pit mine employing 1,100 people. It used some of the world's largest mining equipment, including specially designed 235-tonne capacity, rubber-tired haulage trucks.

In 1980-81, Highmont Mines went into production east of the Lornex pit. Plans were also underway for an even larger mine on the valley floor on claims owned by Valley Copper (Cominco). After hundreds of millions of dollars were spent, Valley Copper did come on line, but with a new name and new partners. Cominco, Lornex and Highmont combined to form Highland Valley Copper. The Lornex mill was expanded by moving the Highmont Mill to the Lornex site. The Valley Copper ore body is now being processed by carrying the ore up to the Lornex mill in giant conveyors.

In 1990, Highland Valley Copper employed 1,220 people to mine 300,000 metric tonnes of ore and waste daily. The 235-tonne trucks were phased out in favor of 170 and 190 tonne capacity haul trucks, supported by huge electric shovels and other heavy equipment.

Highland Valley Road (97C) links the District of Logan Lake with Ashcroft, on the Trans-Canada Highway. This scenic drive climbs gradually into the giant ancient volcanic crater that is the Highland Valley. Approximately 16 kilometres west of Logan Lake, the road passes the former Lornex and Bethlehem Copper operations (now part of Highland Valley Copper) before skirting the growing cavity created by the Valley Copper pit. The road continues the gradual ascent, avoiding the giant tailing pond that collects the mill waste. Then, after reaching the height-of-land, it begins the steep descent into the semi-desert country surrounding Ashcroft.

In the past, Highland Valley Copper has offered mine tours during the summer months. If you can't visit the real thing, a shovel and one of the huge trucks is on display at Logan Lake.

•••

32

Logan Lake Road

Statistics	**For maps, see pages 137 & 140.**
Distance:	53 kilometres, Logan Lake to Kamloops. 24 kilometres to Exit 336 on Coquihalla Highway 5.
Travel Time:	Less than one hour.
Condition:	Paved throughout.
Topo Maps:	Merritt 92 I/SE (1:100,000). Kamloops 92 I/NE (1:100,000).
Forest Maps:	Kamloops Forest District Recreation Map. Merritt Forest District Recreation Map.
Communities:	Logan Lake, Lac Le Jeune and Kamloops.

Logan Lake is an excellent base for outdoor recreation. There is a network of cross-country ski trails on the outskirts of the village that would be the envy of many much larger communities. Follow the ski signs to the parking area and you're set to go. Thirty-six kilometres of trails are groomed by volunteers, with two kilometres of trails lit for night skiing.

For information, write the Highland Valley Outdoor Association, Box 415, Logan Lake, BC V0K 1W0 or call (250) 523-6791.

There is also a network of backroads crossing the plateau to the north, providing access to some prime trout fishing lakes. Tunkwa and Leighton lakes are well-known in the region, but others await you.

Figure 32
This giant electric shovel is part of the Logan Lake Travel InfoCentre.
(Photo LOL-030)

With the crossroads at the Logan Lake information centre as km 0, the highway from Logan Lake to Kamloops skirts the west side of the lake before following the Meadow Creek valley upstream as far as Lac Le Jeune. If fishing is your fancy, you can turn off at km 1.2 for the picnic area and parking lot at the north end of the lake. If you would rather swing at something that doesn't bite or swing back, the Meadow Creek Golf Course is only a few seconds farther along the road.

Figure 33
Shooting at the Coquihalla Outpost sporting clays range. (Photo LOL-049)

For high-country fishing and backroad exploring, head north on Paska Lake Road at km 15.3. Mile High Lodge, on Face Lake, is 11 kilometres north of the highway. It offers cottages and a campground. Dominic Lake Resort is 17 kilometres north of the highway. It is a little higher in altitude although both lakes are close to 1500 metres above sea level. If conditions are right, you can continue on the backroads north of Dominic Lake and emerge on the Trans-Canada Highway, west of Kamloops.

Surrey Lake Forest Service Road at km 16.3 offers you another option for high altitude fishing, backroad exploring and camping. Surrey Lake is about 10 kilometres south of the Logan Lake Road and just east of the summit of the Merritt-Kamloops section of Coquihalla Highway 5. If you welcome mud holes as a challenge, you can try fishing at Sussex Lake, Bob Lake or the Frogmoore Lakes. Make sure you have a winch, as the big one you may want to pull in could be your vehicle — not a wild and wily Kamloops trout.

If you are remotely interested in camping and shooting sports, Coquihalla Outpost Ltd.'s Sporting Clays Resort is worth checking out. Relatively new to the area, this operation is located about 3.5 kilometres off the Logan Lake Road, just west of the Lac Le Jeune Interchange (Exit 336) on Coquihalla Highway 5.

The lay of the land there is perfect for shooting sports. It is rolling

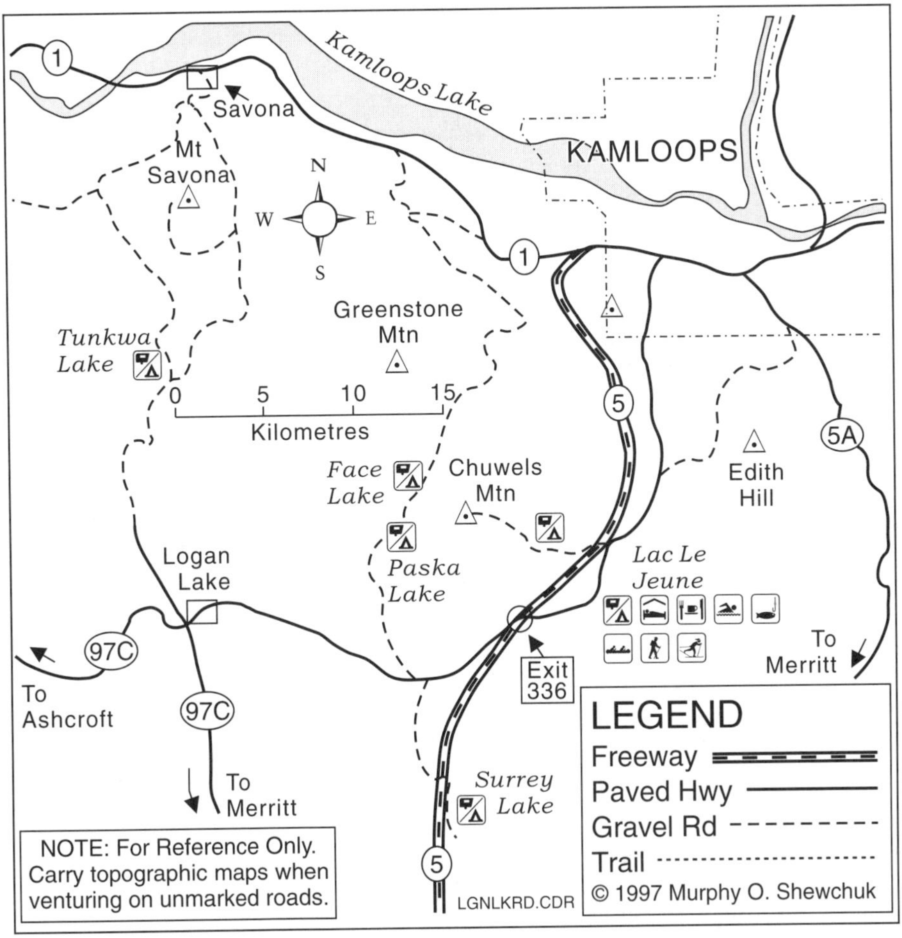

Map 24
Logan Lake — Kamloops area.

plateau country with a beautiful little lake and plenty of lodgepole pine. There are a few old skid roads and some newly built trails through the timber. Log fences have been added and the trees thinned a bit to improve lighting. The sporting clays course has been designed to use the undulating terrain to focus the noise upward and way from other shooters as well as prevent shooters from endangering each other.

The targets at Coquihalla Outpost are various sizes and weights of clay pigeons that are neither particularly predictable nor static. At one station, a launcher high in a tree blind is teamed with one close to the ground and timed to imitate a bird being flushed by the sound of the first shot. At another station, the clays imitate a pair of ducks that suddenly veer off when they see the hunter. Their direction depends on the vagaries of the wind and the clays. Breaking both with one shot is nearly impossible and no simple feat with two.

I did a lot of shooting with a Cooey 22 calibre rifle when I was a teenager, however I have not had much experience with a shotgun and haven't done much shooting since. I tried the circuit, and the real thrill for me was discovering that I could make that shotgun do what I wanted. (Well, quite a bit of the time, anyway.)

The Lac Le Jeune Interchange (Exit 336) is 24 kilometres east of Logan Lake. If you are not in a hurry, take the old route down to Kamloops — you will find plenty of recreational opportunities year around. To make it easier should you be leaving Coquihalla Highway 5 at Exit 336, reset your odometer to zero here and begin the count again.

Lac Le Jeune Road, at km 3.6, provides access to a resort, a now-defunct ski hill, cross-country ski trails, a lake and a provincial park, all under the Lac Le Jeune name. Woody Life Village Resort is 0.6 kilometres off the Logan Lake Road, while the junction to Lac Le Jeune provincial park campground, is another half kilometre farther along. The road straight ahead leads to Lac Le Jeune Resort and the former Lac Le Jeune ski hill. The park and the resort are closed during the winter months. At the time of writing, there were rumors that the park would come under the Discover Camping reservation system. If so, you may have to telephone 1-800-689-9025 for reservations.

With Exit 336 as reference, the junction to the Stake Lake Forest Service Recreation Area is at km 6.5. A cabin near Stake Lake is the base for 45 kilometres of well-groomed cross-country ski trails, managed by the Overlander Ski Club of Kamloops. Chuwels Mountain Road and the Highway 5 underpass, at km 6.6, provides another diversion. It is approximately 11 kilometres to the 1896 metre peak.

Hidden off the highway at km 8.5 is McConnell Lake Provincial Park. Formerly a Forest Service recreation site, little appears to have changed since the change in ownership. Although it doesn't have any services, the fishing can be good. McConnell Lake also marks the beginning of the long descent to Kamloops.

Edith Lake Road at km 13.0, offers another backroad diversion. If luck and conditions are with you, you could end up at Knutsford on Highway 5A, after a pleasant drive through the back country.

North of Edith Lake Road, Logan Lake Road continues its descent, leaving the timber behind and entering the grasslands as it skirts the east side of Sugarloaf Hill and reaches Coquihalla Highway 5 at the Copperhead Drive Interchange (Exit 366) at km 29.

•••

33

Surrey-Sussex Lakes Road

Statistics	For maps, see pages 137 and 140.
Distance:	12 kilometres, Logan Lake Road to Sussex Lake.
Travel Time:	Less than one hour.
Condition:	Gravel road, some rough sections.
Season:	Summer and early fall.
Topo Maps:	Mamit Lake 92 I/7 (1:50,000).
Forest Maps:	Merritt Forest District Recreation Map.
Communities:	Merritt, Logan Lake & Kamloops.

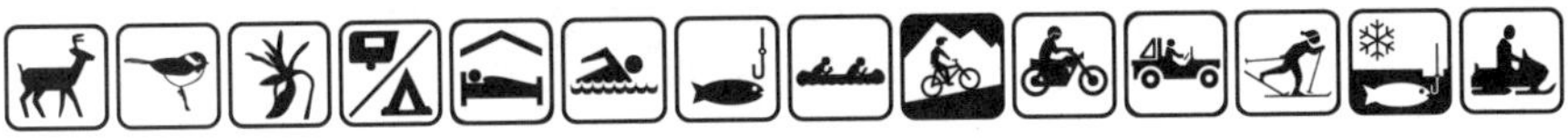

The plateau country southwest of Kamloops has certainly changed since Walter Humphrey established the Surrey Lake Fishing Camp in 1931. Much of the original dirt road from Kamloops has disappeared under the black-top of successive incarnations of Lac Le Jeune Road, Logan Lake Road and Coquihalla Highway 5.

While the Coquihalla Highway passes within hearing distance of Surrey Lake and Sussex Lake, there is no direct connection. Instead, access to Surrey Lake is from Logan Lake Road, 7.5 kilometres west of the Walloper Lake Interchange (Exit 336). The stand of lodgepole pine surrounding the lakes muffle the highway noise, and the nearness of "civilization" is soon forgotten.

With the junction at the Logan Lake Road as km 0, the Surrey-Sussex Lakes Road winds south through a mixture of open fields, lodgepole pine and aspen groves. It crosses Meadow Creek near km 1 and passes the west end of Desmond Lake at km 3.5. Although there is no recreation site or boat ramp at Desmond Lake, it is possible to snake a canoe or car-top boat through the reeds and out into the lake.

Beyond Desmond Lake, the road continues to climb through the pine, passing under the Coquihalla Highway at km 6.9. Keep left at km 7.5, following the main logging road southeast for another half kilometre. The Surrey Lake Fishing Resort sign near km 8 marks the beginning of the rough 1.5 kilometre-long-road into Surrey Lake.

The fishing resort is at the north end of Surrey Lake. The public road, rough and full of pot-holes, continues south along the west side of Surrey Lake, passing the dammed outlet near km 11. It then winds through the trees for about a kilometre to Sussex Lake and the Sussex Lake Forest Service recreation site.

If fishing, camping or back-roads exploring interests you, Surrey and Sussex lakes are certainly worth checking. Unlike some of the marshy upland lakes, Sussex Lake does have a sandy bottom that could be conducive to swimming — if there has been a week or two of hot weather.

In winter, the Surrey-Sussex area could also be a destination for snowmobiling, cross-country skiing and ice fishing.

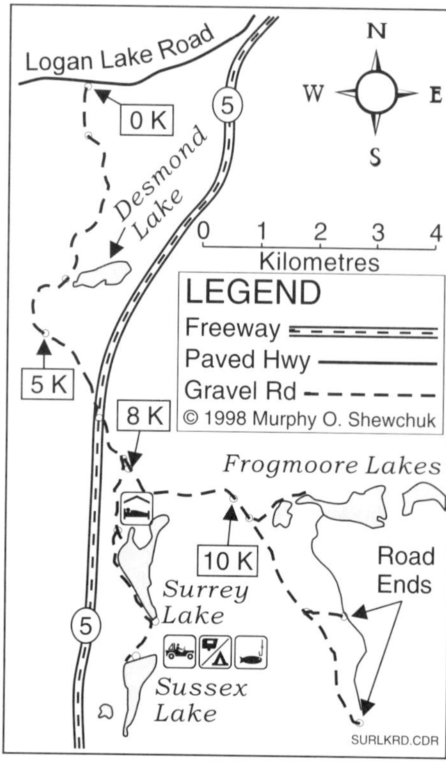

Map 25
Surrey-Sussex Lakes Road

If you prefer the atmosphere of a resort, write to the Surrey Lake Fishing Resort, P.O. Box 489, Kamloops, B.C. V2C 5L2. You can also telephone (250) 523-9778 or e-mail them at <surylake@awinc.com>.

•••

GPS References for major points of interest. See page 9 for details.

Wpt	Description	Latitude	Longitude
W01	0 Km Surrey-Sussex Lakes Rd	N50° 27' 31.486"	W120° 37' 35.837"
W02	Desmond Lake	N50° 25' 47.472"	W120° 37' 54.338"
W03	Highway 5 Underpass	N50° 24' 32.349"	W120° 37' 25.138"
W04	Jct. to Surrey Lake	N50° 24' 10.159"	W120° 37' 03.664"
W05	Surrey Lake Fishing Resort	N50° 23' 31.535"	W120° 37' 08.086"
W06	Sussex Lake FS Rec. Site	N50° 22' 23.403"	W120° 36' 51.690"

•••

34

Chuwels Mountain Road

Statistics	For map, see page 137.
Distance:	12 kilometres, Logan Lake Road to Chuwels Mtn.
Travel Time:	Approximately 1/2 hour in dry weather.
Condition:	Gravel road with rough sections.
Season:	Summer and early fall.
Topo Maps:	Kamloops 92 I/NE (1:100,000).
Forest Maps:	Kamloops Forest Disrtict Recreation Map.
Communities:	Kamloops & Logan Lake.

If you have ever wanted to get to the top, Chuwels Mountain is one place you can do it. This 1896-metre-high timber-clad peak lies north of Lac Le Jeune and south of Kamloops Lake and is easily accessible from the Logan Lake Road. Chuwels Mountain Road starts just north of the Stake Lake recreation site, and after passing under Coquihalla Highway 5, begins its ascent.

The Lodgepole Lake Forest Service rec site near km 4 offers a camping alternative to the provincial park at Lac Le Jeune near the Logan Lake Road. If you are heading to the hills for a day of fishing, you may also want to explore the small lake at km 4.7.

For the next four kilometres, the road winds through logged areas before climbing up through the timber to a fork in the road near km 10.5. The newer road to the right climbs up to the east peak of the mountain and a radio repeater site. The older road to the left continues to the west peak at km 12 and more radio repeaters.

The view from a rocky knoll just west of the buildings on the west peak is excellent. Down below lies Paska and Face lakes while on the middle skyline is Greenstone Mountain. The antenna-festooned mountain in the far west is Mount Savona, another interesting viewpoint that has attracted more than passing interest from local rockhounds.

From the east peak you can see the Knutsford grasslands and the Thompson valley.

•••

Figure 34
Black-capped Chickadee. (Photo BCD-011)

NEW

35

Birdwatching Site Guide

Merritt to Kamloops via Highway 5A

by Rick Howie

Statistics	For map, see page 145.
Distance:	98 kilometres, Merritt to Kamloops via Hwy 5A.
Travel Time:	One to one and one half hours.
Condition:	Paved throughout.
Season:	All season.
Topo Maps:	Merritt 92 I/SE (1:100,000). Kamloops 92 I/NE (1:100,000).
Forest Maps:	Merritt Forest District Recreation Map.
Communities:	Merritt, Quilchena & Kamloops.

Birders travelling north from Merritt to Kamloops have an important decision to make at Exit 290 of the Coquihalla Highway. If you wish to maximize the variety of bird species and birdwatching opportunities, leave the Coquihalla Highway and turn northeast on Highway 5A at the northeastern outskirts of Merritt. Because of the relatively low traffic flow and the productive valley waterways, Highway 5A is a great road to just bird along at an easy pace and stop at any spot that looks interesting. Do, however, drive with caution and park well off the road as heavy transport trucks use this route.

Distances

This special birdwatcher's description of Highway 5A begins at the Coquihalla Highway Exit 290. Set your odometer at this point. Every attempt has been made to synchronize the references with the small white kilometre markers found periodically along the road edge. The features and landmarks described are obvious or well marked. As any odometer error can compound significantly over the 90 kilometre distance, it may be wise to periodically reset your trip meter at a roadside marker. If you follow the sites in sequence, you should have little difficulty finding each one.

Route Guide

Coquihalla Highway Overpass... km 0.0

Swakum Mountain Road... km 1.6

This good gravel road heads north through pasture land, passes under the Coquihalla Highway and climbs up through Ponderosa Pine habitat into the Douglas Fir zone and beyond. The network of roads and loop options goes beyond the scope of this guide but good access to birds typical of the dry interior forests is possible.

Williamson's Sapsuckers have been seen along the lower parts of this road.

Nicola and Mill Creek Road... km 7.9

For the past 6 kilometres, you have been driving through fields of cultivated alfalfa, pastures and other agricultural endeavors. By late May, you might see Bobolink in these fields and Common Snipe on fence posts. You should see Great Blue Herons in flight near the tree-top nesting colony along the Nicola River to the east of the fields. Ospreys also nest on prominent snags close to the Nicola Ranch buildings. In late 1995, Ducks Unlimited Canada improved an old marshy oxbow along the right (southeast) side of the highway for waterfowl. Bald Eagles nest in the adjacent cottonwoods.

A cluster of ranch buildings cleaved by the highway signifies your arrival at the tiny community of Nicola. You might enjoy the aesthetics of the historic Murray Church while sorting out the abundant swallows which typically hawk for insects over the cottonwoods along Mill Creek. Mill Creek Road, on the left, provides access to a few kilometres of riparian habitat before it climbs up through the pines and terminates in a maze of old logging roads. If you are adventuresome and have a high-clearance vehicle, you could connect up with the Swakum Mountain Road and make a loop trip back to Highway 5A. Take the wrong turnoff however, and several seasons could pass before another lost birder finds you. The Mill Creek Road is also worth exploring for Pacific-slope Flycatchers that can be found in the cottonwoods during the summer.

At the northern edge of Nicola (km 8.3), the Monck Park Road departs to the left and provides access to good Ponderosa Pine habitat and the west side of Nicola Lake. The fields along this road have

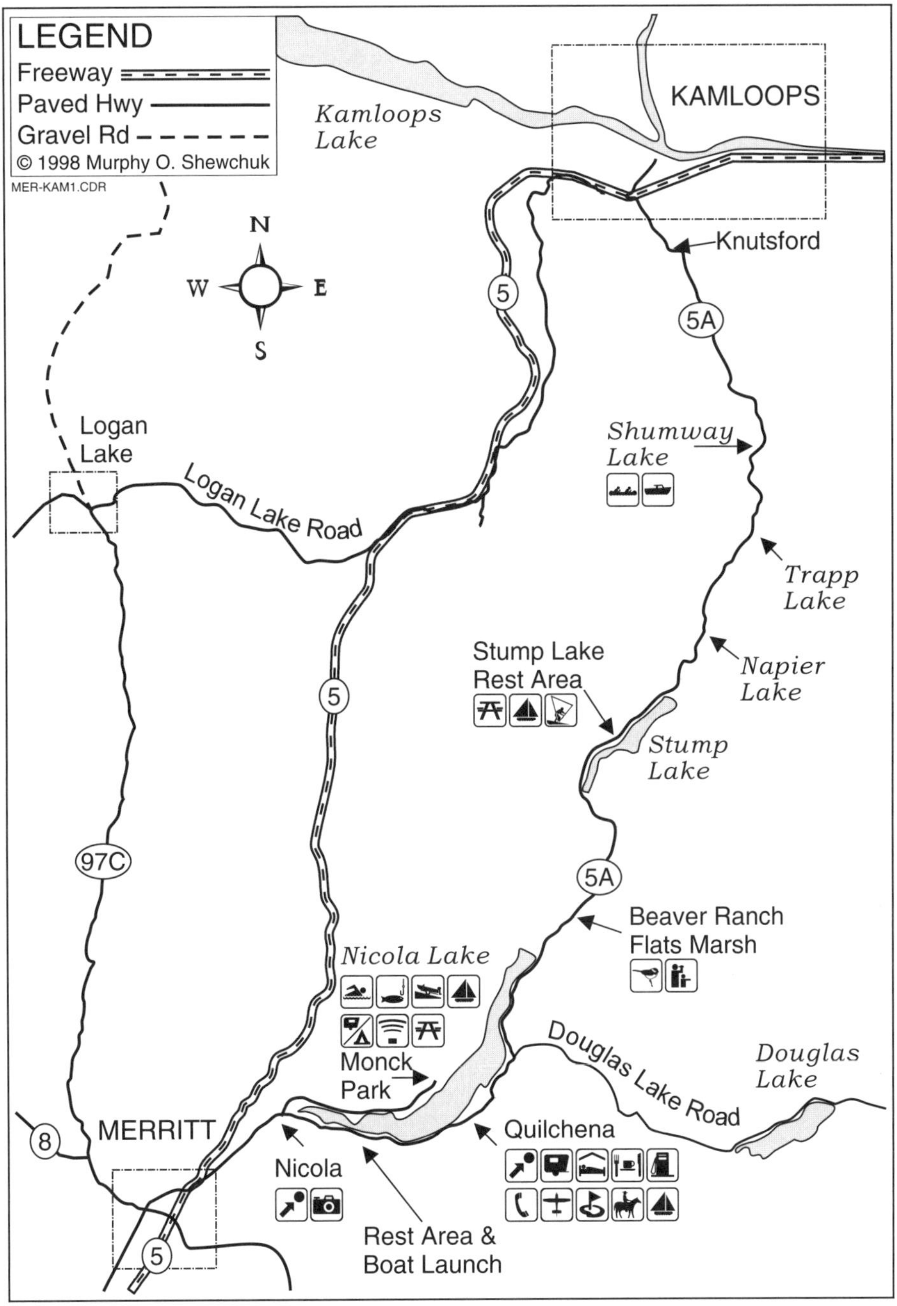

Map 26
Merritt to Kamloops via Highway 5A.
Note: Full services are available in Merritt and Kamloops. Some services at Quilchena are seasonal and others may have limited hours.

Figure 35
Redwing Blackbirds frequent the cattail marshes along Highway 5A.
Murphy Shewchuk photo.

produced Bobolink. Willow Flycatchers are common in the shrubs along the lake. Golden Eagles have nested on the cliffs to the west of the road. House Wrens congregate in the small housing community along the lake edge as well as along the park access road. Chukars occur in small numbers above the cabin complex and Flammulated Owls have been reported in Monck Park.

Back at Nicola, continue north along Highway 5A and cross the Nicola River that drains Nicola Lake. A dirt road departs on the right and runs obtusely back in a southerly direction towards Merritt. It crosses grasslands of the Merritt Commonage and provides access to Sharptailed Grouse habitat. These birds dance and display on their leks during the early morning hours from March until late May. Note that they are usually gone from the dancing sites by about 8:00 a.m.

Proceeding north along the main highway takes you past the productive south end of Nicola Lake. Waterfowl can be abundant during the spring and fall migration periods but viewing distances are a little long. Watch for Osprey that nest on the power poles and on the hillsides to the east of the highway. You may also see Tundra and Trumpeter Swans as well as White Pelicans on this portion of the lake.

Cattail Marsh... km 11.0

This marsh can be productive for Marsh Wrens, American Bittern, Virginia Rail, Sora, Yellow-headed Blackbird and Willow Flycatcher.

Rest Area... km 12.6

This is a convenient spot if the last comfort break that you had was somewhere back on the Coquihalla. You can scan the lake from this point and it is often productive for waterfowl in migration, particularly during the late fall. In summer, study the cliffs and broken, shrubby terrain along the east side of the highway, just past the rest stop. This terrain is home to large numbers of Cliff Swallows and a scattering of House Wrens, Nashville Warblers and Lazuli Buntings.

Marina Viewpoint... km 19.4

As part of the 1993 Summer Games, a marina was built here as a legacy facility. The shallow bay to the north of the parking lot is used for loafing by a variety of birds, especially when lake levels are low and sand bars are exposed. September can be productive for White Pelicans, gulls and Common Terns. Bonaparte's Gull can be common during the first two weeks of May. This is also a good time to watch for Western Grebe, Surf Scoter, White-winged Scoter and other waterfowl.

Quilchena Hotel... km 20.1

A trip up the Nicola Valley is not complete without a stop at this historical landmark. The food is great and the golf course deceptive. Great Horned Owls can be found in the trees around the hotel or heard along the nearby lake shore. A Mockingbird was found on the hotel roof one winter. However, more normal summer birds include Veeries and Northern Waterthrush along the lake edge, Savannah Sparrows, Cliff and Barn Swallows and possibly Bobolink. Vesper Sparrows are found in the drier areas behind the hotel. Kokanee spawn in the creek crossing the golf course during September and skeins of Sandhill Cranes pass overhead in April and again in September. Large flocks of Canada Geese often occupy the pastures across the road and north from the hotel during migration. Several pairs of Osprey nest in the vicinity.

Pennask Lake Road... km 22.1

For a truly rewarding exploration of the grassland environment, this road is a must. Depending upon the season, you can experience impressive raptor migrations that include Swainson's, Rough-legged and Red-tailed Hawk, Northern Harrier, Kestrel and Prairie Falcon. The extremely rare Ferruginous Hawk has been seen on occasion. Other species of interest in the appropriate season include American Avocet, Sharptailed Grouse, Long-billed Curlew, Burrowing Owl (very rare), Short-eared Owl, Long-eared Owl, Horned Lark, Clay-coloured Sparrow, and Snow Bunting. The movements of sparrows during the spring and fall are usually quite heavy with Chipping, Vesper, Savannah, Lincoln's, and White-crowned sparrows predominating.

Willow-lined slough... km 22.5

Just past the Pennask Lake Road turnoff lies a small slough on the left separated from Nicola Lake. Pulling off can be tricky as the highway curves in this area. Parking back at the Pennask Lake turnoff and walking would be a safer alternative. You may see a variety of dabbling ducks and Canada and Ross's geese here. During the late spring and fall migrations, the muddy edge may harbour shorebirds. Both species of yellowlegs, Long-billed Dowitcher, Baird's Sandpiper, Least Sandpiper and Red Phalarope have all been seen here. Tree Sparrows often rest here in the late fall.

Douglas Lake Road... km 25.0

Marked by an attractive log church, this turnoff takes you through the famous Douglas Lake Ranch and ultimately leads to Hwy 97 at Westwold and on to the Okanagan. Birdwatching can be excellent across the Douglas Plateau despite the restricted access.

Nicola River... km 26.5

The highway crosses the delta of the Nicola River with pullouts nearby if you look carefully. Watch the treetops, for Osprey nest on snags here. During September, Herring, California, and Ring-billed Gulls loaf on the sand bar while snacking on Kokanee. Watch the lake for Common Loons, grebes, both species of goldeneye and other waterfowl. Cliff Swallows nest under the bridge and occasionally, a Dipper can be found wintering along the open stream.

Guichon Ranch... km 32.9

The ponds at the entrance to the ranch can be very productive for waterfowl in the spring and fall. Both dabbling and diving ducks use the ponds that are close to the road and provide good viewing. Three species of teal, Northern Shoveler, Redhead, Ring-necked Duck, scaup, goldeneye and Ruddy Duck are typical. Long-billed Curlews sometimes use the wet seepage in the fields to the south of the ranch entrance road. Watch for Lewis' Woodpeckers at Ponderosa Pine trees as you drive north from here to the next stop.

Beaver Ranch Flats... km 35.3

This wetland enhancement project was finished in 1990 by Ducks Unlimited Canada in order to benefit waterfowl and other marsh-oriented species. In cooperation with the Guichon family, this managed area has become a virtual Mecca for birds both in migration and during the breeding season. In addition to a long list of ducks, the shallow ponds have hosted White Pelican, American Avocet, Baird's Sandpiper, Least Sandpiper, Long-billed Dowitcher and Sora and Black Tern. Eared and Pied-billed Grebes nest here while Bald Eagles are frequently seen during late February and March. Merlins chase shorebirds here during early September. A fenced pullout provides safe viewing well off the highway. Remember, this is private land!

Peterhope Road... km 41.5

A large sign advertising Peter Hope Lake Resort marks this turnoff which leads to upper elevation fir and pine forests. For the first kilometre or so you may be able to flush Ruffed Grouse from the riparian thickets along the right hand side of the road. The shrub thickets along Highway 5A just south of this junction are home to Long-eared Owls.

Planet Mine Road... km 42.7

Turn off to the right for access to Ponderosa Pine and Douglas Fir habitats. Typical species include Clark's Nutcrackers, three species of nuthatches, Mountain Bluebirds, Townsend's Solitaire, Mountain Chickadee and many other passerine species.

Stump Lake south... km 44.1

This pullout provides some safe views of the productive south end of Stump Lake. In addition to large numbers of ducks and coots, watch for Black Swifts and Black Terns over the lake, especially if there is a large hatch of midges during the summer. Yellow-headed Blackbirds are common here. For about the next eight kilometres, you will be travelling near the shore of Stump Lake. There are a few opportunities to pull off and scan for waterfowl, including Pied-billed, Red-necked and Eared Grebes as well as a variety of passerines. Watch for nesting Northern Orioles and Eastern Kingbirds.

In the winter, Sharptailed Grouse pick buds in the tops of cottonwoods while Short-eared Owls patrol the adjacent grasslands.

Watch for White Pelicans from mid to late April and large numbers of Sandhill Cranes around April 20.

About five kilometres to the north, a rest stop with toilets provides good access to the lake edge. Water birds are not usually too abundant at this popular rest stop, but do scan the lake anyway. You may see diving birds such as Surf and White-winged scoters.

Tullee Lake Marshes... km 54.8

Ducks Unlimited Canada has worked with Stump Lake Ranch to enhance this area for waterfowl. Tullee Lake itself is closer to the road at km 55.1. Although small, it is often a good spot to see Red-winged Blackbirds as well as Canvasback, Redhead, scaup, Bufflehead, Goldeneye, Gadwall and other ducks.

Napier Lake Viewpoint... km 57.9

A large boulder and Stop of Interest sign entitled "Empire of Grass" marks this pullout above Napier Lake. Check the bluffs on the opposite side of the road for Golden Eagles that nest here regularly.

Depending upon the time of year, the lake below may have good numbers of diving ducks, but viewing conditions are not the best for close views. Across the highway and just a bit to the south, a marshy depression is often excellent for waterfowl, nesting Canada Geese and migrant passerines. Do not be surprised to see exotic waterfowl as the owners raise a variety of birds in this private marsh.

Napier Lake north... km 59.6

A shallow marshy area separates Napier Lake from Ritchie Lake. In the latter half of May, spawning suckers attract small concentrations of Bald Eagles and Common Mergansers. As pullout conditions are poor, continue to the next stop for safer viewing of Ritchie Lake.

Ritchie Lake pullout... km 60.5

A widening on the left side of the road provides a safe pullout to view Ritchie Lake. This is the most productive of the chain of lakes that you will pass on the route north from this point. Diving and dabbling ducks can be abundant during the spring and fall migration months. During chironomid (midge) hatches, large numbers of swallows swoop and dart over the lake.

Trapp Lake south... km 62.3

Pullouts are awkward for much of the nine kilometre length of Trapp Lake. However, except for the south end, it seldom shows many bird concentrations. Grebes and diving ducks are the most abundant water birds. The pullout at the south end of the lake can be one of the best spots to watch Black Terns during chironomid hatches in late May. Goldeneye and scaup may form small flocks here as well.

Trapp Lake north... km 65.5

Ducks Unlimited Canada enhanced the marshes near the north end of the lake. You can often enjoy excellent views of dabbling ducks such as Northern Shoveler, Mallards and all three species of teal.

Roche Lake Road... km 67.0

Eleven kilometres of good gravel road leads to Roche Lake Resort. Accommodation is excellent as is the fishing and birdwatching. A Forest Service recreation site is also located at Roche Lake. Birds of interest include Barred, Great Horned, Great Gray and Saw-whet Owls. Townsend's Warblers abound in the Spruce woods. The wail of Common Loons adds to the romance of this upper-elevation lake.

Shumway Lake... km 70.0

Pull off the road here beside the south end of Shumway Lake. There is usually little in the way of birds north of the boat launch site, part way up the lake. A pair of Bald Eagles has nested on the opposite side of the lake just north of the boat launch site. Common Mergansers can be "common" at times.

Campbell Creek Road... km 74.0

The road that leaves Highway 5A to the right provides an interesting circuitous tour into Kamloops. It passes through some good forested birding habitats that include stands of Douglas Fir, Ponderosa Pine and Black Cottonwood. If you want a change of pace and do not mind adding 25 kilometres to your trip, take this side road. Turn left when you reach pavement again some 10 km along and you will eventually reach Kamloops via a well-signed route. As you continue along Highway 5A, the road climbs Cardew Hill where Golden Eagles occasionally ply the skyline.

Figure 36
Eagles can be frequently seen near the lakes along Highway 5A.
Murphy Shewchuk photo.

Separation Road... km 77.4

This road departs to the right and climbs into the grasslands for a few kilometres before terminating at a private ranch corral. Buildings along this road often harbour Say's Phoebes. During April, you may surprise a flock of Sandhill Cranes. You may also find such typical grassland birds as Horned Larks and Vesper Sparrows or even Lapland Longspurs during September. At the start of this road, the wetland usually has good numbers of noisy Yellow-headed Blackbirds.

Jackson Road... km 77.6

A left turn puts you onto Jackson Road which climbs over the hills to connect with Long Lake Road. Long Lake Road will, in turn, bring you back to Highway 5A at km 85.5. This optional loop provides opportunities to see House Wrens in aspen-lined gullies, Sharptailed Grouse, Mountain Bluebirds and a host of other species. However, most of the land is privately owned and permission from the ranchers must be obtained before wandering from the road. The Haughton family are owners of much of the property along Jackson Road but do not live along it. Some members live along Separation Road, if you wish to seek information about hiking on their land.

Beresford Road... km 80.5

Continuing north along Highway 5A from Jackson Road, you pass

Separation Lake, in a depression along the right side of the highway. Beresford Road departs to the right and appears to lead into a private ranch house. In fact it does, but you can drive straight ahead at the ranch driveway and head eastward across the grasslands on this public road right-of-way. Close all wire gates if you proceed through any of them. This is cattle country and the ranchers do not appreciate wandering stock. This road does provide access to Separation Lake and good grassland birdwatching, but private land abuts all sides of the road. Do not leave the road without permission. As you proceed north along the highway from this point, watch over the fields and drumlin on the right side of the road. This entire area is excellent for migrating raptors in the spring and fall. You may also see Sandhill Cranes during April and hordes of waterfowl on Separation Lake and the two ponds just to the south of it.

Rosehill Road... km 84.0

Rosehill Road provides excellent grassland birding opportunities and can be a great place to watch migrating raptors. The opportunities are too complex to describe here, but if you follow the road, it will ultimately take you to Kamloops. It can be worth the drive.

Knutsford... km 85.5

You have reached the small settlement of Knutsford and a junction to the left to Long Lake Road. As with the previous side road, good birding opportunities exist along a circuitous trip back into Kamloops. Do not be afraid to explore. You can always backtrack. The rangeland to the west of the highway is home to Sharptailed Grouse, Horned Lark, and Great Gray Owl. Roads climb up through Douglas Fir forest and lakes such as Edith and Goose can be productive. Williamson Sapsuckers have been seen near the upper end of Goose Lake Road.

Kamloops... km 90.0

Trans Canada Highway 1 and the friendly city of Kamloops marks the end of this productive route. More terrific birding opportunities abound in and around the city!

I hope that you have enjoyed your birdwatching trip from Merritt to Kamloops. May I recommend that you return south along the same route. It is almost impossible to see all of the birds or explore all of the side roads during one trip.

Good Birding! Rick Howie

•••

36

Monck Park

Statistics	For map, see page 155.
Distance:	11.2 kilometres, Highway 5A at Nicola to Monck Park.
Travel Time:	One quarter hour.
Condition:	Paved throughout.
Season:	Accessible year around.
Topo Maps:	Merritt 92 I/SE (1:100,000).
Forest Maps:	Merritt Forest District Recreational Map.
Communities:	Merritt.

Monck Provincial Park, 22 kilometres northeast of downtown Merritt, is more than just a lake-side campground.

Set among the ponderosa pines on the north shore of Nicola Lake, it was once considered to be just an overnight stop on the way to some distant exotic destination. However, with the increasing number of people traveling the highways, "exotic destinations" aren't what they used to be. As a result of the growing interest in low impact recreation and the natural environment, local and out-of-town visitors are turning the park into a popular "destination." In addition to the obvious water activities such as swimming, boating and water-skiing, there is the opportunity to angle for the more than 20 species of fish that are known to inhabit Nicola Lake.

Some even hint that Nicola Lake may be deep enough to harbor its own version of the Ogopogo or Loch Ness Monster. However, unfounded hints of something bigger than a kokanee or rainbow trout have not stopped water sport lovers from sailing and wind surfing most summer afternoons.

Off the water, the semi-arid environment attracts more than people. The scattered pines shelter a wide variety of wildflowers, birds and small animals. If you stop to listen, you may hear an owl high in a tree or a chipmunk rustling among the dead branches near your feet. The song of the coyote is not uncommon, although it is usually much farther away than it seems.

Be careful where you sit while listening to the natural world — you may discover a prickly pear cactus hidden among the bunchgrass.

Figure 37
Lava beds near the trail above Monck Park. (Photo MON-063)

Monck Park Trails

For those interested in slightly less passive nature observation, approximately three kilometres of trails have been developed along the waterfront and on the hillside above the park. The lakeshore trails converge at the picnic area and beach and generally wind between the campground and the water. The hillside trails climb up from the waterfront and through or above an ancient volcanic outcropping. They also cut across old creek beds and the arid open slopes. With each change in micro-climate there is a corresponding change in vegetation and wildlife. The variety of plants, animals and birds and the relatively easy trail makes this an ideal walk for patient parents with younger, curious children — just be sure to pack along a snack and a water jug.

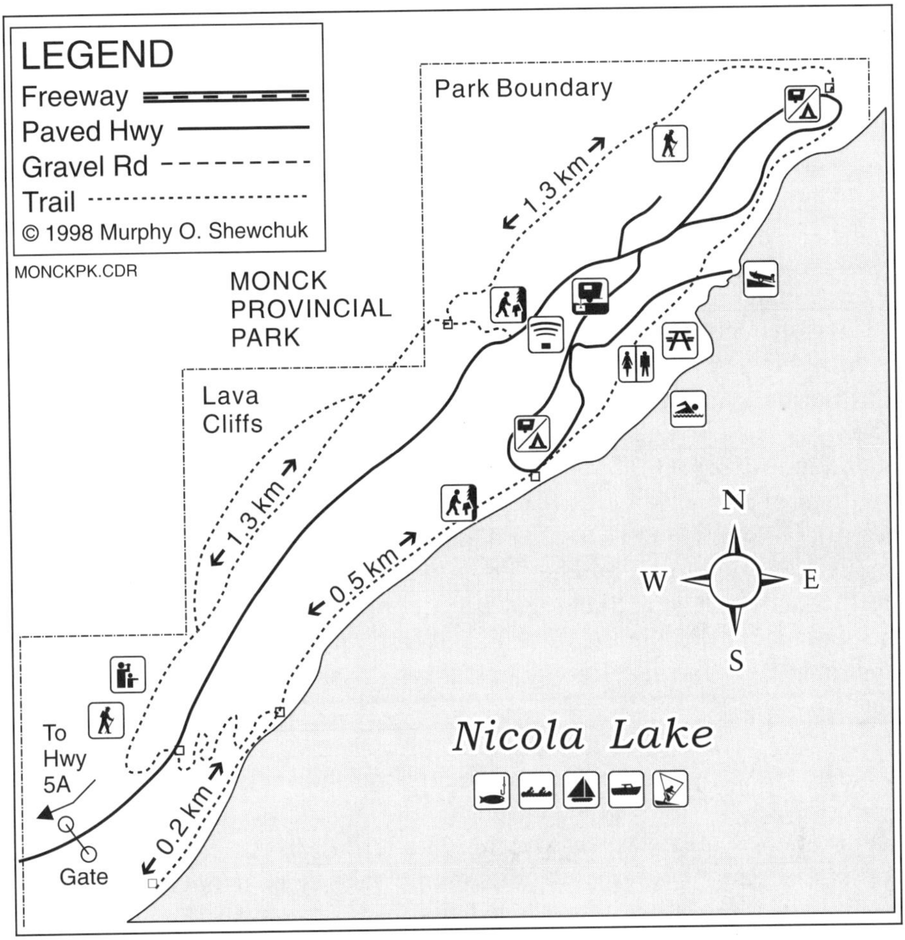

Map 27
Monck Provincial Park

The near-view is not the only point of interest to be seen from the park trails. Across the lake sits the historic Quilchena Hotel. If you look closely at the grassland backdrop to the hotel, you may see the shoreline benches of the glacial lakes that filled the valley some 10,000 years ago.

Although not part of the park, there is also an old road/hiking trail along the lake shore northeast of the park. It leads to an area known as Second Beach and a fine example of Indian rock paintings (pictographs). Access is over or through the fence at the northeast boundary of the park. The four-kilometre hike to Second Beach could take an hour each way, but the opportunities to photograph wildlife and dry-country wildflowers could make it worth the effort. This land had been vacant and undeveloped for decades, but it was sold to new owners in 1994. At the time of writing, a lakeshore housing development has been proposed for the property so future access is uncertain.

Eighty-seven hectare Monck Provincial Park was created in 1951 on land set aside to protect it from just such development. Major Goldman donated the land that served as the nucleus for Monck Provincial Park in honour of his son, Lieutenant-Commander Penryn "Pen" Monck, an officer in the British Navy during World War II. Penryn Monck officially changed his name to Monck (an old family name) in 1941, some say to improve his chances of survival should his ship be captured by the German fleet.

Charles Sydney Goldman was born in Bungherdorp, South Africa in 1863. At an early age he was a partner in a large gold and diamond firm at Johannesburg. Goldman later served as a Member of Parliament in England before coming to the Nicola Valley in 1919. Shortly after his arrival, Goldman purchased 6,550 acres of land situated between Merritt and Nicola, including much of the townsite of Nicola and a power dam on the Nicola River. Starting with a couple of hundred cattle, Major Goldman built the Nicola Stock Farms Ltd. up to over 5,000 head. Major Goldman, and his wife Lady Agnes, lived in the Nicola Valley until his death on April 3, 1958.

Access is via the 11.2-kilometre-long Monck Park Road which leaves Highway 5A at the community of Nicola, approximately 10 kilometres northeast of Merritt. In addition to the campground and trails, the park is equipped with picnic tables, boat launching ramp, drinking water, a landscaped beach area and change houses.

Monck Park is also an excellent summer base for exploring the rest of the Nicola Valley. In addition to programs by the Park Naturalist, the park draws on outside resource people to present a varied perspective of the region.

For more information on Monck Park, contact the BC Parks District Office in Kamloops at (250) 851-3000. At the time of writing, there are rumors that the park would come under the Discover Camping reservation system. If so, you may have to telephone 1-800-689-9025 for reservations.

•••

37

Pennask Lake Road

Nicola Lake to Highway 97C

Statistics	**For maps, see pages 160 & 100.**
Distance:	55 kilometres, Quilchena to Bear Road junction.
Travel Time:	Two to three hours in good weather.
Condition:	Gravel, some rough sections.
Season:	June to October.
Topo Maps:	Douglas Lake 92 I/1(1:50,000). Paradise Lake 92 H/16 (1:50,000). Peachland 82 E/13 (1:50,000).
Forest Maps:	Merritt Forest District Recreation Map.
Communities:	Merritt, Kelowna, Westbank & Peachland.

Pennask Lake, on the Douglas Plateau southeast of Quilchena and west of Kelowna, has a history that is both unusual and colorful. Among the more famous personalities to spend time at this upland lake are the American Pineapple King, James Dole, and Queen Elizabeth II and Prince Philip of England.

The first rough road into Pennask Lake wound eastward from Quilchena, via Minnie Lake through the grasslands and the mud holes. Getting there was no easy task, even in the best of weather. And the worst of weather could mean days stuck in the quagmire. In the late 1960s, the opening of the Brenda Mines copper/molybdenum property helped provide easier access to Pennask Lake from Peachland. Construction of a 138 kV power line from Nicola Substation to Brenda Mine and, in 1980-81, a 500 kV line to the Kootenays opened a 93-kilometre-long dry-weather road from Quilchena to Peachland.

Although open to the public, it is not currently being maintained by the Ministry of Highways. Instead, BC Hydro and logging companies maintain it when necessary. Trucks and smaller recreational vehicles do make the trip during the dry summer months, but not without their fair share of bumps and dust.

Access to the west end of Pennask Lake Road is off Highway 5A, approximately two kilometres north of the Quilchena Hotel (25 kilometres northeast of Merritt). The road climbs steadily for the first 10 kilometres before levelling out on the Douglas Plateau.

After leaving Nicola Lake, the first visible signs of man's activities are a group of corrals and buildings nestled near a small lake in a stand of aspen and cottonwood, 14 kilometres southeast of Highway 5A. The Quilchena Dry Farm was established in 1913 as a provincial government experiment to determine if the natural grasslands surrounding the Nicola Valley were suitable for homesteading. For eight years, government-employed farmers dealt with an annual rainfall averaging 25 centimetres; a maximum of three frost-free months; and temperatures ranging from -30° to +35° Celsius. Then, in 1920, just as a good crop of oats was maturing, a plague of grasshoppers destroyed the entire year's work. The government immediately abandoned the experiment and the land reverted to cattle range.

Minnie Lake, near km 20, was given its name by Byron Earnshaw after his daughter Minnie. Earnshaw and his Indian wife, Shinshinelks, settled in the area in the 1870s. Brian K. de P. Chance, a long-time manager of the Douglas Lake Ranch, first proposed that Minnie Lake be declared a waterfowl sanctuary in 1939. Since the 1941-42 season, many thousands of ducks, geese, swans and pelicans have rested unmolested on their twice-annual fly-past.

In more recent years, Douglas Lake Ranch has constructed a beautiful lodge on nearby Stoney Lake.

A sign at the junction near km 22 marks a side road that winds 16 kilometres northward and down to Douglas Lake and the Spahomin Indian Reserve. See *Douglas Lake Road* on page 161 for details.

The road to the right continues eastward to Pennask Lake and Peachland, passing through the Flume Marsh soon after leaving the junction. Several dams on Wasley Creek create or upgrade a total of 75 hectares of marshland as part of Ducks Unlimited Canada's first project on the 220,000 hectare Douglas Lake Ranch. The Flume Marsh, the lowest of the Wasley Creek projects, was so named because of the nearby flumes that once carried creek water to the Minnie Lake Ranch hay fields. The dams now control the water to the benefit of wildlife and waterfowl, and pumps drive the giant irrigation sprinklers.

The former Paradise Lake Road branches southward near the 28K signpost. There are nine lakes in the immediate area (elevation 1,500 metres) with eight accessible by good trails, creating a fly fisherman's dreamland. Access has been restricted from Pennask Lake Road and is now from the Elkhart Interchange on Highway 97C (see page 99).

The road again forks near km 33. Pennask Forest Service Road, to the left, was the original road to Pennask Lake Lodge. While there is easier access to Pennask Lake via Bear Road, considerable work is being done on a series of hiking, cycling and snowmobiling trails in the Wasley/Mellin/Holmes lakes area. When complete, the trails will link Pennask and Hatheume lakes as well as provide shorter hikes to several lakes on the plateau. Plans were underway to produce a brochure that would be available through the Forest Service office in Merritt.

The road to the right dips down across Quilchena Creek and generally follows the power line right-of-way. There are several side roads along the route, but one near the 43K marker leads to Reservoir Lake, Skunk Lake and Highway 97C at the Sunset Main Interchange. There is a Forest Service recreation site at Reservoir Lake "with very rough four wheel drive access". (See *Paradise Lakes Area*, page 99.)

A junction at the east end of the right-of-way road, approximately 50 km from Highway 5A, is of major importance. After looping under the highway bridge across Pennask Creek, it reaches a pair of junctions. At the first, the choice is to go right to the Sunset Main Interchange. Left leads to the junction of Bear Road and Sunset Main Road. Bear Road passes under the highway via a giant culvert before continuing past Pennask Lake and Hatheume Lake and, eventually, to Okanagan Lake near Bear Creek Provincial Park, opposite Kelowna. Sunset Main Road winds east and south past the Headwaters Lakes to Peachland.

•••

GPS References for major points of interest. See page 9 for details.

Wpt.	Description	Latitude	Longitude
W01	Jct. Highway 5A & Pennask Rd	N50° 10' 16.536"	W120° 29' 29.484"
W02	Douglas Lake Ranch Dry Farm	N50° 04' 09.282"	W120° 26' 18.024"
W03	Roadside Marshes with Ducks	N50° 01' 55.416"	W120° 25' 17.388"
W04	Minnie Lake Ranch Gate	N50° 01' 27.894"	W120° 24' 29.088"
W05	Jct. to Douglas Lake	N50° 01' 54.312"	W120° 22' 48.894"
W06	24K - Flumes	N50° 01' 07.368"	W120° 22' 17.688"
W07	Paradise Lake Road	N49° 59' 41.196"	W120° 20' 26.316"
W08	Jct. 0K Pennask FS Rd	N50° 00' 16.074"	W120° 17' 04.140"
W09	Jct. to South near 42K Marker	N49° 57' 04.362"	W120° 12' 05.388"
W10	43K Marker & Jct. to Reservoir Lk	N49° 56' 40.146"	W120° 11' 10.056"
W11	Jct. Reservoir Lk to Right	N49° 56' 00.096"	W120° 11' 59.688"
W12	Reservoir Lake Rec. Site	N49° 55' 14.634"	W120° 12' 19.368"
W13	Sunset Main Interchange	N49° 53' 58.140"	W120° 09' 59.010"
W14	Bear Road/Pennask Cr. Overpass	N49° 54' 24.054"	W120° 05' 32.676"

•••

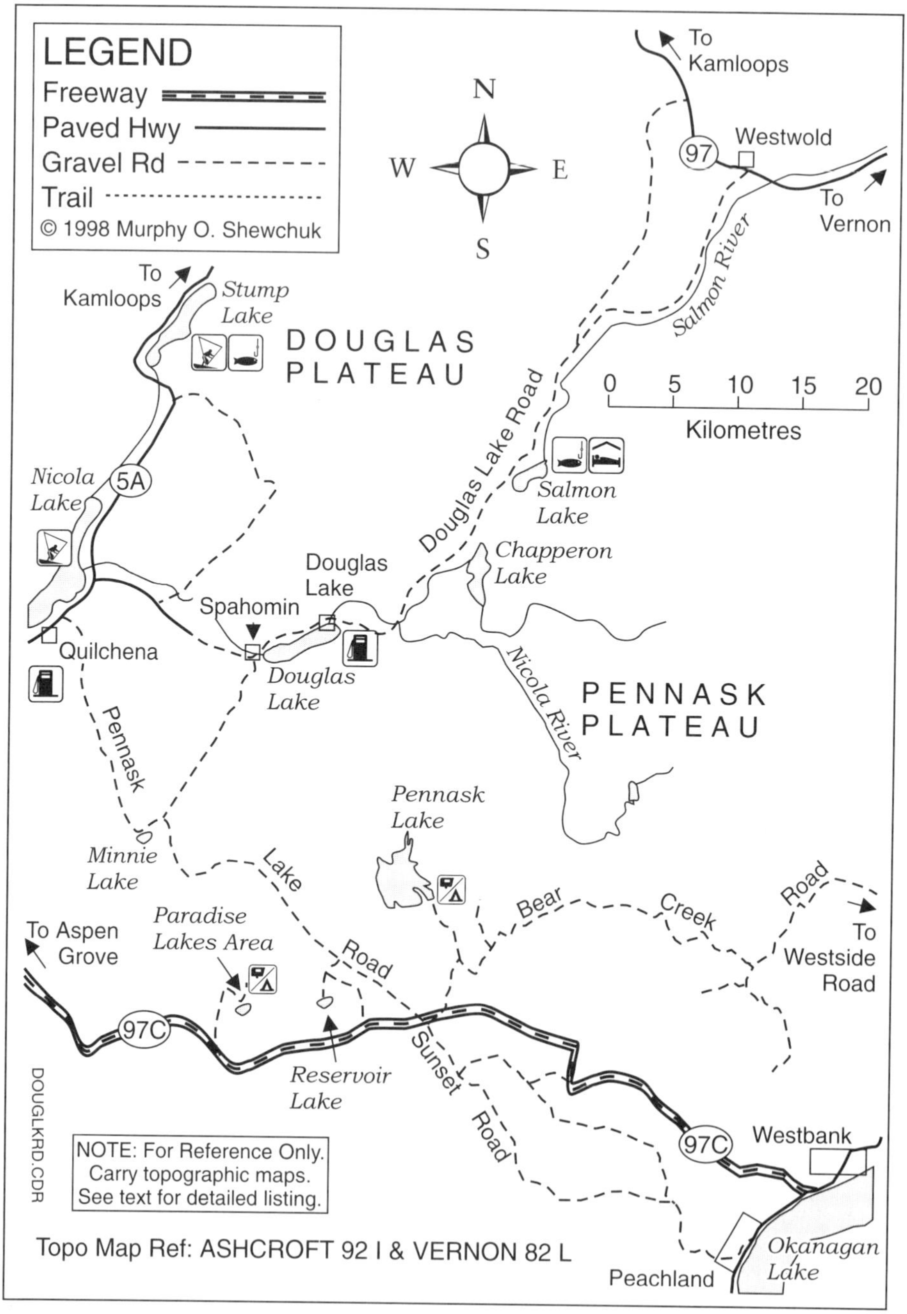

Map 28
Douglas Lake Road: Nicola Lake to Westwold.

38

Douglas Lake Road

Statistics	For map, see page 160.
Distance:	80 kilometres, Nicola Lake to Westwold.
Travel Time:	One to 1.5 hours.
Condition:	Mostly gravel with paved sections.
Season:	All season.
Topo Maps:	Merritt 92 I/SE (1:100,000). Vernon 82 L/SW (1:100,000).
Forest Maps:	Merritt Forest District Recreation Map.
Communities:	Merritt, Quilchena, Douglas Lake & Westwold.

In the upper reaches of British Columbia's Nicola Valley lies an empire of bunchgrass, sagebrush, timber and lakes half the size of Canada's province of Prince Edward Island. Bisecting this rich cattle country is a meandering 80-kilometre-long backroad that passes through biotic zones ranging from semi-desert grasslands to cedar forest.

The Douglas Lake Road, gravel for most of its length, leaves Highway 5A at the Quilchena Indian Reserve, 28 kilometres north of Merritt and 68 kilometres south of Kamloops. At the west end, this route snakes along the lower slopes of a broad valley. Irrigated hay fields cover the alluvial fan near the Nicola River mouth while prickly pear cactus, sagebrush and bunchgrass dot the arid slopes. A twisting line of cottonwood trees mark the meandering Nicola River.

Across the valley, a barely visible horizontal line, best seen at sunrise or when the first winter snows dust the rangeland, defines the shoreline of an ancient glacial lake. During the last period of glaciation, which ended some 10,000 years ago, this region was ripped open and reformed by the flowing mass of destruction. As the ice retreated, it created three different glacial lakes.

Glacial Lake Quilchena was the first major lake to form in the Nicola Basin. It reached an elevation of 1,000 metres and drained southward into Otter Creek near Aspen Grove. The second, Glacial Lake Hamilton, formed when the ice plug retreated in the Salmon River Valley, permitting eastward drainage through the Douglas Plateau. Some of Lake Hamilton's shoreline terraces are still visible on the north slope of Hamilton Mountain, southeast of Nicola Lake.

The last of the giant glacial lakes, Lake Merritt, was formed when the ice melted out of the Campbell Creek Valley and allowed the meltwater to flow northward into the Thompson Valley near Kamloops. As the last ice plug melted out of the lower reaches of the Nicola Valley, it opened the present drainage route via Spences Bridge. Glacial Lake Merritt left the shoreline now visible from the Douglas Lake Road.

Near the Lauder Road junction (km 7.2), more than a kilometre of steel link fence encloses Nicola Substation, one of B.C. Hydro's largest substations.

An ancient weather-beaten cabin sits on the north bank of the Nicola River, approximately three kilometres up Lauder Road. This is known as the Beak Cabin. According to Nina Woolliams in her book, *Cattle Ranch*, Charles Miles Beak had a varied and colorful career as a cook, gold miner, cattle drover, butcher and rancher before settling in the Nicola Valley in 1878. In 1883, Beak began a concerted land and cattle purchasing program that quickly made him one of the largest ranchers in the Interior. Later that same year, he formed a partnership with J.B. Greaves that resulted in the Douglas Lake Ranch.

Lauder Ranch, a short distance farther up Lauder Road, has the distinction of being one of the oldest ranches in the Nicola Valley that is still owned by the founding family. Joseph Dixon Lauder established the ranch in 1876 and it is still operated by his great-grandson John. (See *Glimpse Lake Loop*, page 167, for more information on Lauder Road and Glimpse Lake, 13.5 km to the north.)

Hamilton Mountain Road is marked by a pole gate at the crest of the hill overlooking Douglas Lake, near km 12.5. The 5.6 kilometre drive (or hike) affords an excellent view of Douglas Lake and the surrounding grasslands from several natural viewpoints. The broad panorama of the open rangeland is even more easily seen from the 1,500 metre peak. (See *Hamilton Mountain Road*, page 173.)

The Douglas Lake Road crosses the Nicola River at the outflow of Douglas Lake, in the Native community of Spahomin. The community derives its name from the willows that grow along the shore of Douglas Lake. The young branches of the willows were shaved by the Okanagan Indians and used in the construction of dip nets. The name apparently refers to the many piles of shavings that resulted. Another historic landmark at Spahomin is the Catholic Church that is situated well off the road in the Native village. A side road leads south from Spahomin, into the heart of the Douglas Lake Ranch, joining the Pennask Lake Road near Minnie Lake. See *Pennask Lake Road*, page 157, for more information. Though a little rough, it is usually passable by car in dry weather and can be followed back to Highway 5A near the historic Quilchena Hotel.

Figure 38
Horses on a grassland hillside near Douglas Lake Road.
(Photo 252-B-11)

At the east end of Douglas Lake (km 25) lies the community of Douglas Lake, the headquarters of the Douglas Lake Cattle Company. Formerly owned by the late Charles "Chunky" Woodward, this 222,750 hectare ranch is considered the largest in Canada.

John Douglas, a Scottish born pioneer of no relation to British Columbia's first Governor, Sir James Douglas, first settled the bottom-land near the east end of the lake in 1872. His holdings became part of the nucleus of this vast operation when Charles Miles Beak and "Old Danger" Greaves recognized the potential market in the construction crews of the Canadian Pacific Railway. With the help of a handful of Victoria businessmen, Greaves formed the Douglas Lake Cattle Company. From a foundation built on the lucrative railway contracts, the ranch grew until today it is a multi-million dollar operation. With a staff of close to 50 people, the ranch ships approximately 7,000 white-faced Herefords annually and winters at least 10,000 cattle on hay and grain harvested from their own fields.

East of the ranch headquarters, on the crest of the hill overlooking Sanctuary Lake, is the ranch's private airstrip, capable of handling small jet traffic. The rail-fenced corrals at English Bridge, a short drive further east, is the home of the Panorama cattle sales held each September. Once the auctioneers begin their lively routine, it usually takes less than an hour for up to 3,500 Douglas Lake prime yearling steers to change hands.

Beyond this last bridge over the Nicola River, the gravel road continues eastward through the open rangeland past Chapperon and Salmon lakes. The group of buildings on the east shore of Chapperon Lake form the headquarters for the eastern operation of the Douglas Lake Ranch. In the early spring, the fields and barns become home to hundreds of new calves while the lakes and marshes provide a resting place for ducks, geese and sandhill cranes on their annual spring migration northward.

Chapperon Lake is also one of the first plateau lakes to open each spring. For this reason, the Okanagan Indians valued the lake and established a fishing camp here in the days long before the influence of the white settlers.

Salmon Lake, 48 kilometres from Nicola Lake on the Douglas Lake Road, is one of several lakes in the Nicola Valley that are restricted to angling with artificial flies. Salmon Lake, like many of the lakes and streams in the region, also has a restricted angling season.

Salmon River drains eastward, eventually emptying into Shuswap Lake at the city of Salmon Arm. For nearly 10 kilometres, it winds through the plateau ranchland, then it descends into a narrow canyon. The Douglas Lake Road follows this same route, leaving the dry grasslands behind and entering the Interior wet belt near km 51. Bunchgrass and ponderosa pine give way to lodgepole pine, Douglas fir and cedar.

The volcanic rock bluffs along the roadway near km 70, 10 to 11 kilometres west of Westwold, are a favorite haunt of local rockhounds in search of moss and banded agate nodules.

Figure 39
Rockhounds explore the bluffs near Westwold. (Photo RHO-011)

Westwold was originally named Grande Prairie by the nineteenth century fur traders and the Westwold area was often used as a camp for the fur brigades on their way from Kamloops to the Columbia River, via the Okanagan. Westwold also has the dubious distinction of being the final home of the last surviving camel imported to pack supplies during the Cariboo Gold Rush of the 1860s. The camel story began in 1862 when John C. Callbraith of Seton Portage, near Lillooet, purchased 23 bactrian camels at $300 each. Callbraith saw the twinkle of gold when he read that the camels could carry 275 kilograms of freight and live off the grass and shrubs along the Cariboo road. As with most get-rich-quick schemes, this one had a flaw. The camel's terrible smell frightened livestock and their voracious appetite included everything from sagebrush to miner's socks. In addition to their social unacceptability, their feet could not stand up to the sharp rock on the mountain trails, and the camels were soon turned loose to fend for themselves.

The last of the camels survived until about 1905, spending his final years as a ranch pet in the Westwold area.

Today, Westwold, and the surrounding Salmon Valley, is part of the Shuswap-North Okanagan dairy country, taking advantage of the rich soil and mild climate recognized long ago by the fur traders.

There is always talk of upgrading the Douglas Lake road, usually heard loudest around provincial election time, and surveyors occasionally appear — and disappear — with little left to show of their passing except the stakes and colorful ribbons. Talk and rumours, however, have been heard in cattle country for more than a century with little obvious change in the ranching way of life.

•••

GPS References for major points of interest. See page 9 for details.

Wpt	Description	Latitude	Longitude
W01	Hwy 5A - Douglas Lake Rd	N50° 11' 32.457"	W120° 28' 20.272"
W02	Lauder road	N50° 10' 45.297"	W120° 23' 38.338"
W03	Hamilton Mountain Road	N50° 09' 02.538"	W120° 21' 03.379"
W04	Spahomin Bridge	N50° 08' 23.296"	W120° 17' 01.439"
W05	Prince Philip's Point	N50° 09' 29.421"	W120° 14' 00.197"
W06	Douglas Lake 111 Sign	N50° 09' 57.345"	W120° 13' 13.887"
W07	Douglas Lake Headquarters	N50° 09' 58.523"	W120° 11' 52.739"
W08	Douglas Lake Airfield	N50° 09' 59.972"	W120° 10' 35.221"
W09	Nicola River at English Bridge	N50° 10' 03.815"	W120° 08' 35.912"
W10	End of pavement	N50° 11' 25.330"	W120° 07' 25.249"
W11	Chapperon Lake view	N50° 12' 57.293"	W120° 03' 43.684"
W12	Rush Lake	N50° 15' 10.855"	W120° 02' 30.453"
W13	Salmon Lake Resort	N50° 16' 28.354"	W120° 00' 12.779"
W14	Norfolk Ranch	N50° 18' 14.704"	W119° 59' 16.484"
W15	Junction - keep right	N50° 20' 05.670"	W119° 58' 09.086"
W16	Top Salmon River canyon	N50° 20' 21.158"	W119° 57' 30.346"
W17	Salmon River crossing	N50° 21' 38.870"	W119° 55' 43.011"
W18	Highway 97 - Westwold	N50° 28' 26.177"	W119° 46' 57.205"

•••

39

Glimpse Lake Loop

Statistics	For maps, see pages 168 and 160.
Distance:	34 kilometres, Douglas Lake Road to Hwy 5A.
Travel Time:	One to two hours in dry weather.
Condition:	Gravel road with rough sections.
Season:	Best in summer and early fall.
Topo Maps:	Merritt 92 I/SE (1:100,000).
Forest Maps:	Merritt Forest District Recreation Map.
Communities:	Merritt, Quilchena & Kamloops.

This backroads drive from the Douglas Lake Road, near BC Hydro's Nicola Substation, to Highway 5A near Stump Lake offers upland trout fishing, excellent panoramic views and a glimpse at the Nicola Valley's colorful past.

The gravel road begins as Lauder Road, 7.2 kilometres east of Nicola Lake on the Douglas Lake Road. It initially skirts the Nicola Substation and then winds down to the Nicola River at km 2.0. The cottonwood trees near the river provide welcome shade on a hot day and the back-eddies in the stream harbour some fine pan-size trout.

An ancient cabin sits on the north bank of the Nicola River at km 2.6. The Douglas Lake Cattle Company's sign identifies it as the Beak Cabin. According to Nina Woolliams in her book, *Cattle Ranch*, Charles Miles Beak had a varied and colorful career as a cook, gold miner, cattle drover, butcher and rancher before settling in the Nicola Valley in 1878. In 1883, Beak began a concerted land and cattle purchasing program that quickly made his one of the largest ranches in the B.C. Interior. Later that same year, he formed a partnership with J.B. Greaves that was to establish the land base for the giant Douglas Lake Ranch.

John Lauder's "Spring Bank Ranch" (JL), located at km 4.8, has the distinction of being one of the oldest in the Nicola Valley still owned by the founding family. Joseph Dixon Lauder established the ranch in 1876 and it is still operated by his great-grandson John.

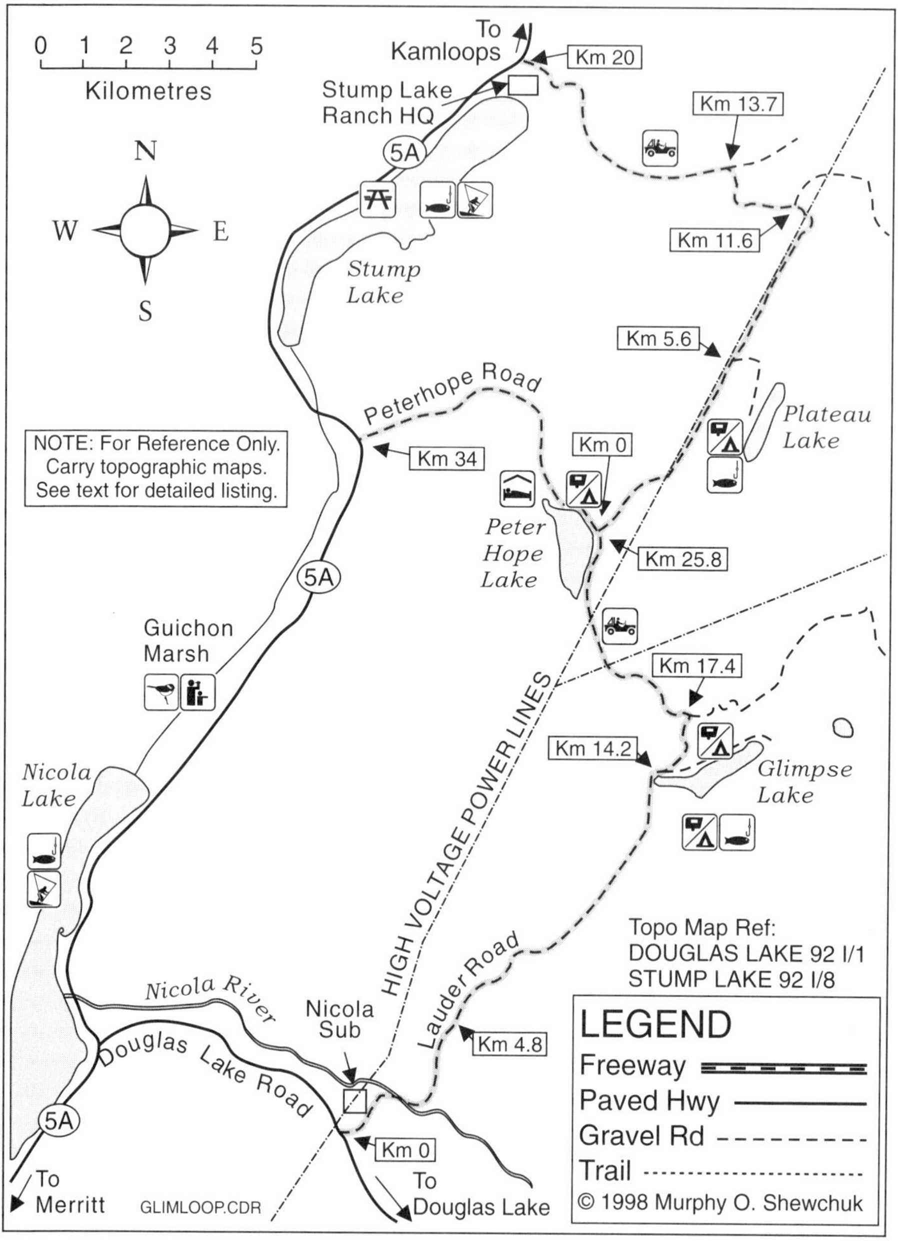

Map 29
Nicola Lake - Peter Hope Lake - Stump Lake Area.

A handsome log cabin set near the edge of Lauder Creek at km 6.8 has been the site of several movies. Known as the "Abbott Place", Joe Lauder purchased the home and ranch from Hamilton Abbott in the mid-1950s. The house served as an office for the CBC production, *Red*

Serge, a light comedy series based on the Northwest Mounted Police that was filmed nearby in 1985 and 1986. The building was also part of the set for the film "*One Boy, One Wolf, One Summer*" undertaken by Gem Productions in 1988.

After passing the Lauder Ranch and the Abbott Place, the road continues to climb steadily through the grasslands, winding in and out of old creek beds. A long-abandoned cabin and corral at km 9.0 mark the past while cattle on the range and a beaver swimming in a nearby slough are the only signs of present life.

One of the many problems faced by Nicola Valley ranchers is soil erosion triggered by man and beast. A hillside near the cattleguard at km 11 shows the signs of motorbike activity. Some of the ruts in the dust have been carved much deeper by the spring snow melt and frequent summer thundershowers that can dump a considerable amount of rain in a matter of minutes.

The road leaves the grasslands at km 12.5 and enters the timber just before reaching the south end of Glimpse Lake at km 13.5. There is a small Forest Service Recreation Site near the south shore of the lake as well as a boat launch area on the lake. There are private homes along the lake shore and in the trees as well as a second small Forest Service recreation site on the north shore of Glimpse Lake.

A junction at km 14.2 marks the first major option since leaving the Douglas Lake Road. The road straight ahead passes a few more homes and a Forest Service recreation site before ending at the Beaver Creek Ranch. The road to the left (west) continues through the subdivision. What has been a good, well-maintained gravel road to this point now becomes somewhat narrower and certainly lumpier as it continues to climb northwest into the timber.

Keep left at the junctions at km 16.6 and km 17.2. The rough cross-road ends on a much better logging road at km 17.4. The road to the east follows the BC Hydro right-of-way toward Monte Lake and may not be passable. Peter Hope Lake lies to the west and this route is usually in much better condition. The route peaks near km 18 at an elevation of approximately 1300 metres having climbed about 525 metres since leaving the Douglas Lake Road. It now begins a steady descent, passing under high voltage electric lines at km 22.2 and 23.5.

The "10K" sign at km 23.7 marks the southeast end of Peter Hope Lake and the distance from Highway 5A near Stump Lake. There is an opening in the trees near this end of the lake that could be a good place to park while fishing among the reeds. There are also several homes along the east shore of the lake. A junction near the "8K" signpost (km 25.8) marks the beginning of a very rough loop road into Plateau Lake. (See *Plateau Lake Loop*, page 171, for details.)

Figure 40
Some "serious" fishing on Glimpse Lake. (Photo 419-B-12)

Continue north along Peter Hope Lake to the large Forest Service Recreation Site. Its open layout makes this one of the better recreation sites in the region. Beyond the outlet of the lake, the road begins a steady descent to Highway 5A, passing the entrance road to the Peter Hope Lake Resort near km 27.5. The 400 metre descent in seven kilometres is made a little more difficult by the washboard nature of the gravel road. (Or did the steep grade contribute to the washboard?) Use caution on the turns and narrow sections. Watch for the Peter Hope Lake Resort or "Peterhope Road" signs when approaching this 34-kilometre-long backroads loop from Highway 5A.

•••

GPS References for major points of interest. See page 9 for details.

Wpt	Description	Latitude	Longitude
W01	Km 0 Lauder Rd	N50° 10' 42.863"	W120° 23' 29.570"
W02	Beak Cabin	N50° 10' 52.423"	W120° 21' 57.703"
W03	Km 4.8 Lauder Ranch HQ	N50° 11' 54.646"	W120° 21' 14.290"
W04	Km 14.2 Glimpse Lake	N50° 14' 51.350"	W120° 17' 27.955"
W05	16K Marker Peterhope Rd	N50° 15' 47.972"	W120° 15' 23.470"
W06	8K Marker - Road to North	N50° 17' 55.528"	W120° 18' 38.733"
W07	Peterhope Lake FS Rec. Site	N50° 18' 04.662"	W120° 19' 07.643"
W08	Hwy 5A Peterhope Rd	N50° 18' 56.128"	W120° 23' 02.456"

•••

40

Plateau Lake Loop

Statistics	**For map, see page 168.**
Distance:	20 kilometres, Peter Hope Lake to Hwy 5A. 28 kilometres, Hwy 5A (Peterhope Rd) to Hwy 5A at Stump Lake Ranch.
Travel Time:	One to two hours in dry weather.
Condition:	Gravel road with very rough sections.
Season:	Summer and early fall.
Topo Maps:	Merritt 92 I/SE (1:100,000).
Forest Maps:	Merritt Forest District.
Communities:	Merritt, Quilchena & Kamloops.

This backroads jaunt into the high country northeast of Merritt has very few redeeming features except for isolation, lots of mud holes and a fine upland lake. If you are out to take your kids for a Sunday drive in the old Chevy, there are safer places to explore. But if the "old Chevy" happens to be a 4x4 pickup with lug tires and lots of clearance *and* the kids happen to be particularly adept with a chain saw, shovel, winch and fly rod this could be your choice of destination.

The south end of this backroad leaves Peterhope Rd at the 8K sign on the north shore of Peter Hope Lake. Peterhope Rd joins Highway 5A just south of Stump Lake on the old Kamloops-Merritt Highway. (See *Glimpse Lake Loop*, page 167, for additional details.)

With the 8K signpost at Peter Hope Lake as the km 0 reference, the rough gravel road runs north, paralleling the high voltage lines although still in the timber.

At km 3.2, the road emerges from the timber and crosses under the 500 kV lines. At this point the road changes from rough to terrible. For the next 2.4 kilometres it skirts mud holes, climbs washed-out slopes and generally tries to make things difficult for man and beast. It was not, however, impassable at the time of writing. A motorhome pulling a cartop boat on a trailer also made it through.

A junction in the middle of the BC Hydro right-of-way at km 5.6 marks the spur road to Plateau Lake. To the right the narrow one-lane cow path winds two kilometres through mud holes and over loose gravel to a beautiful little Forest Service recreation site on the shores of Plateau Lake. Officially designated as a six-unit site, it has provisions for day-use visitors as well as a cut through the reeds that can serve to launch your canoe or cartop boat.

Meanwhile, back at the km 5.6 junction.

If you don't want to face the grueling road from Peter Hope Lake again, you can head north along the right-of-way for about five kilometres to km 11. This section has fewer rocky hills but more potential mud holes, so the choice depends on the season. In spring and early summer, it may be safer to head back to Peter Hope Lake.

Assuming you have made it to km 11, follow the road west across the right-of-way and into the lodgepole pine. There are incremental improvements in the road as it is joined by forest roads coming in from the right at km 11.6 and 13.7. Keep straight ahead at both junctions and you should leave the timber and enter the grasslands high above the Stump Lake Ranch at km 16.3.

Keep to the main road as it descends through the grasslands. If you want to stop to enjoy the view of Stump Lake and the ranching scenes below, pull well off the road in order to avoid conflict with ranch equipment or logging trucks.

The Plateau Lake Loop ends (km 20) at Highway 5A near the Stump Lake Ranch gate just north of Stump Lake and approximately 57 kilometres north of Merritt. The complete backroad loop is about 28 kilometres from Highway 5A back to Highway 5A and will take about two hours under ideal conditions. A serviceable, well-equipped vehicle is a must if you are to avoid a long walk out.

•••

41

Hamilton Mountain Road

Statistics	**For map, see page 174.**
Distance:	5.6 km, Douglas Lake Rd to Hamilton Mtn Lookout.
Travel Time:	Less than 1/2 hour.
Elevation Gain:	600 metres.
Condition:	Gravel road with rough sections.
Season:	April through October.
Topo Maps:	Merritt 92 I/SE (1:100,000). Douglas Lake 92I/1 (1:50,000).
Forest Maps:	Merritt Forest District Recreation Map.
Communities:	Merritt, Quilchena & Kamloops.

If you are looking for an excellent view of the grasslands of the Douglas Plateau or an opportunity to photograph fairy slippers and chocolate lilies, consider this short backroad.

Figure 41
Calypso bulbosa. (Photo WOC-083)

Hamilton Mountain Road leaves Douglas Lake Road at the crest of a hill 13 kilometres east of Highway 5A and Nicola Lake. It climbs through the grasslands for 1.4 kilometres before entering the light timber that cloaks the dry slopes of the mountain. Cattle and horses often range on these slopes, providing a foreground for a "wild-west" panorama of rolling grasslands and distant mountains.

There is a gate at km 2.2 that marks the boundary of the Hamilton Commonage. Just beyond the gate, at km 2.5, a sheltered rock bluff to the right of the road is an ideal spot to photograph fairy slippers (Calypso bul-

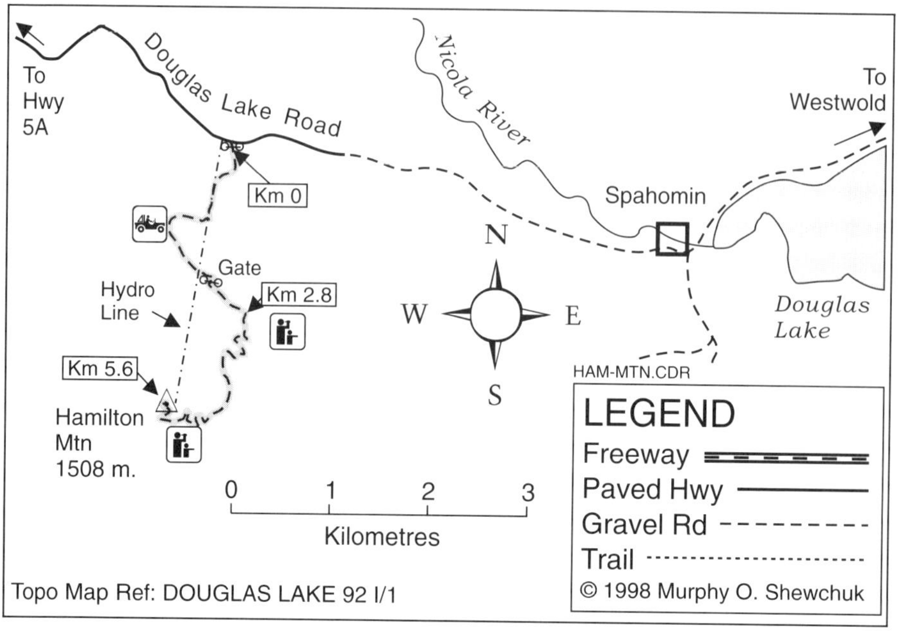

Map 30
Hamilton Mountain Road.

bosa) and chocolate lilies in May and early June.

The narrow dirt road continues to wind around the 1500 metre high mountain, reaching the Forest Service Lookout and a microwave radio site at km 5.6. The broad panorama visible from the catwalk of the fire lookout can be particularly interesting. Study the slopes near Douglas Lake and you will see the shorelines of the giant glacial lakes that inundated the valley one hundred centuries ago.

Although rough in a few places, the Hamilton Mountain Road is usually passable with two-wheel-drive cars or light trucks in dry weather. However, I would certainly not recommend it for motorhomes or vehicles pulling trailers.

•••

GPS References for major points of interest. See page 9 for details.

Wpt	Description	Latitude	Longitude
W01	Km 0 at Douglas Lake Road	N50° 09' 01.882"	W120° 21' 02.259"
W02	Junction near power line	N50° 08' 36.448"	W120° 21' 15.584"
W03	Second gate	N50° 08' 16.498"	W120° 21' 11.201"
W04	Km 2.8 - Viewpoint	N50° 08' 06.727"	W120° 20' 53.588"
W05	Km 5.6 - Hamilton Mtn Summit	N50° 07' 36.987"	W120° 21' 33.911"

•••

NEW

42

Chataway-Pimainus Loop

Statistics	For map, see page 179.
Distance:	45 kilometres, Lower Nicola to Pimainus Lake. 27 kilometres, Pimainus Lake to Shackan (Hwy 8).
Travel time:	One half to one day.
Elevation gain:	Approximately 1100 metres.
Conditions:	Partly paved, mostly gravel with rough sections.
Season:	Best in dry weather. Winter access to Chataway Lake Resort.
Topo maps:	Merritt 92 I/2 (1:50,000). Mamit Lake 92 I/7 (1:50,000). Spences Bridge 92 I/6 (1:50,000).
Forest maps:	Merritt Forest District Recreation Map.
Communities:	Merritt, Lower Nicola & Spences Bridge.

The Chataway Lake area, despite the efforts of Bill Roddy, the Willson family and others, has remained relatively unknown to all except the serious angler. The dozen lakes that make up the Chataway group are not famous for particularly large fish, nor are they particularly large themselves. The greatest attraction is the solitude in which a fisherman can cast a fly or drag a lure through the water. Despite the fact that Coquihalla Highway 5 passes through Merritt, 40 road kilometres away, it is still possible to have a lake to yourself, even on a summer Saturday afternoon. And it's still possible to reel in wild rainbow trout when lower elevation lakes are slow.

One of the reasons for Chataway's isolation was the road. Actually, "wagon trail" was probably a better description until Aspen Planers Ltd. of Merritt began harvesting some of the lodgepole pine that grow out of the rocky highlands. They carved a new road into the area, using much of the old road, but bypassing some of the muddier and steeper sections. This has made it possible to get into the area with little difficulty.

Figure 42
A fisherman on Chataway Lake.

The Chataway Lake adventure starts at Lower Nicola, 8.5 kilometres west of Merritt on Highway 8. Kilometre 0 is the junction of Aberdeen Road and Highway 8. Follow Aberdeen Road north for 3.3 kilometres and the relentless climb to the highlands begins.

The Craigmont copper mine, the first major landmark to the west, originally began in 1960 as an open-pit mine. When the open pit had gone as deep as it could safely go, the miners moved underground.

By the time Craigmont closed in 1983, the mine had produced over 426 million kilograms of copper, worth over $450 million. All that remains are a few mine buildings and a flat field behind the tailing dam. This "field" is presently being re-processed for magnetite powder, a Craigmont "waste" product that is now used in a coal-cleaning slurry in some of BC and Alberta's larger coal mines.

"Chataway" signs mark the junction at km 6.5, a few hundred metres from the mine gate. The road to the right climbs up a draw, passing the local shooting range. To minimize confusion (mine and yours), it is probably simpler to reset your odometer to zero at the Craigmont junction and use the Forest Service kilometre markers as a reference.

Stumbles Road, at 4K on the markers, is the next major intersection. H. Stumbles was a mining man who worked the Eric claim on the Promontory Hills in the 1930s. Although his diamond-drilling efforts missed the Craigmont ore body, he did uncover copper and iron in the form of chalcopyrite and magnetite. Stumbles Road to the left (west) provides access to the Promontory Hills while the main road continues north to the 7K marker where it again forks. A "Chataway" sign marks the road to the left as the best route to the plateau country.

If you are interested in mines or prospecting, you may want to detour six kilometers up the old road to the right in search of remnants of the Aberdeen Mine. Mineral showings were first reported at the Aberdeen Mine site on Broom Creek as early as 1897. Copper was mined at infrequent intervals over the next three decades until 1928 when a major mining effort resulted in gold, silver and over 177,000 kilograms of copper. Efforts were made in the 1960s to reopen the property, but excessive water forced its closure.

Meanwhile, back at the 7K junction, a turn up the hill begins a climb northwest up the Skuhun-Pimainus Forest Service Road. The Coquihalla Highway, this is not. With a realistic speed of 50 to 60 kilometres per hour, everyone but the driver can enjoy the upland plateau scenery. Tyner Lake, everybody's secret fishing hole, lies in a hollow to the west of the road between 14K and 15K. If you are lucky enough to find the correct backroad into the Tyner Lake Forest Service Recreation Site, it can also become your secret fishing hole. (Clue: Look for a "No. 4 Rd" marker.)

The climbing eases considerably around 15K, and the road winds through stands of lodgepole pine and around upland swamps and ponds. Keep right at the Tyner-Skuhun junction at 15K. Most of the junctions are marked with "Chataway" signs, some of which are painted on broken paddles or pieces of plank.

If you are interested in some serious backroad exploring, Tyner-Skuhun Road offers an opportunity to explore a few more lakes, but I wouldn't recommend doing it with the family sedan. There is a rough road into Abbott Lake that starts about one kilometre west of the main road. According to the maps, it also continues south past Gordon Lake and down to Highway 8.

Skuhun Road, near 19K, is one of the original routes to the Chataway area. A right turn here will take you up to Chataway Lakes Resort. For the next six kilometres, the road twists and turns up Chataway Creek. A junction about 3.7 kilometres from the Skuhun-Pimainus Road will lead you east across several clear-cuts to Dot Lake and then north to the Antler Lake Trail or south to Gypsum Lake and a hiking trail up Gypsum Mountain. (See *Chataway Trails*, page 181, for details.)

A second junction a few hundred metres before the Chataway Lakes Resort gates is another decision-making point. To right (south east) is the old road to Dot and Gypsum Lakes. Straight ahead is Chataway Lakes Resort and to the northwest is the old road to Roscoe Lake.

The Willsons' Chataway Lakes Resort offers clean log cabins, a campground and boat rentals. Aside from a recreation room and a few basic supplies, the rest is up to you. Bring your own food, bedding and anything else you might want, including enough fuel to get you back out again. Chataway Lake, at 1310 metres elevation, stays relatively cool year around. According to fishing experts, Chataway and the other lakes in the immediate area are self-sustaining with enough moving water to support proper spawning.

Roscoe Lake, lies six kilometres northwest of Chataway Lake over a narrow, rough road. At 1585 metres, it is near the top of the plateau. There is a recreation site and hiking trails in the area. (See *Chataway Trails*, page 181, for details.)

Frequent logging roads and mining exploration survey lines bisect the plateau. The highlands form the southern ridge of the Guichon Batholith, a giant crater-like porphyry deposit now mined by Highland Valley Copper, one of the world's largest copper mines.

Meanwhile, back on the Skuhun-Pimainus Road, Skuhun Creek Road heads west near the 19K marker. Before construction of Skuhun-Pimainus Road, the Skuhun Creek Road from Highway 8 at Shackan (14 Mile) was the better way into the area. The road is rough with some sandy sections. Should you decide to take this way out of the Chataway Lakes area, you may notice the distinct ecological zones. By the time the road crosses Skuhun Creek, the lodgepole pine stands begin to thin out, replaced by interior Douglas fir and ponderosa pine. Where the road parallels the creek bed, the growth is lush and green. But the semi-desert climate is soon apparent where it crosses the benches. Scattered ponderosa pine, kinnikinnick and bunchgrass are all that survive the hot summer winds that blow up the valley.

Bev Veale's Kinnikinic Homestead, eight kilometres from Highway 8, is one of the few visible signs of human habitation along the route. Bev offers rustic housekeeping cabins and a campground with the emphasis on backcountry vacations. She even has corrals for your horse.

Beyond the 19K marker, Skuhun-Pimainus Road continues a steady climb northward to the heart of the plateau. If you are interested in still another diversion, watch for a junction just past the

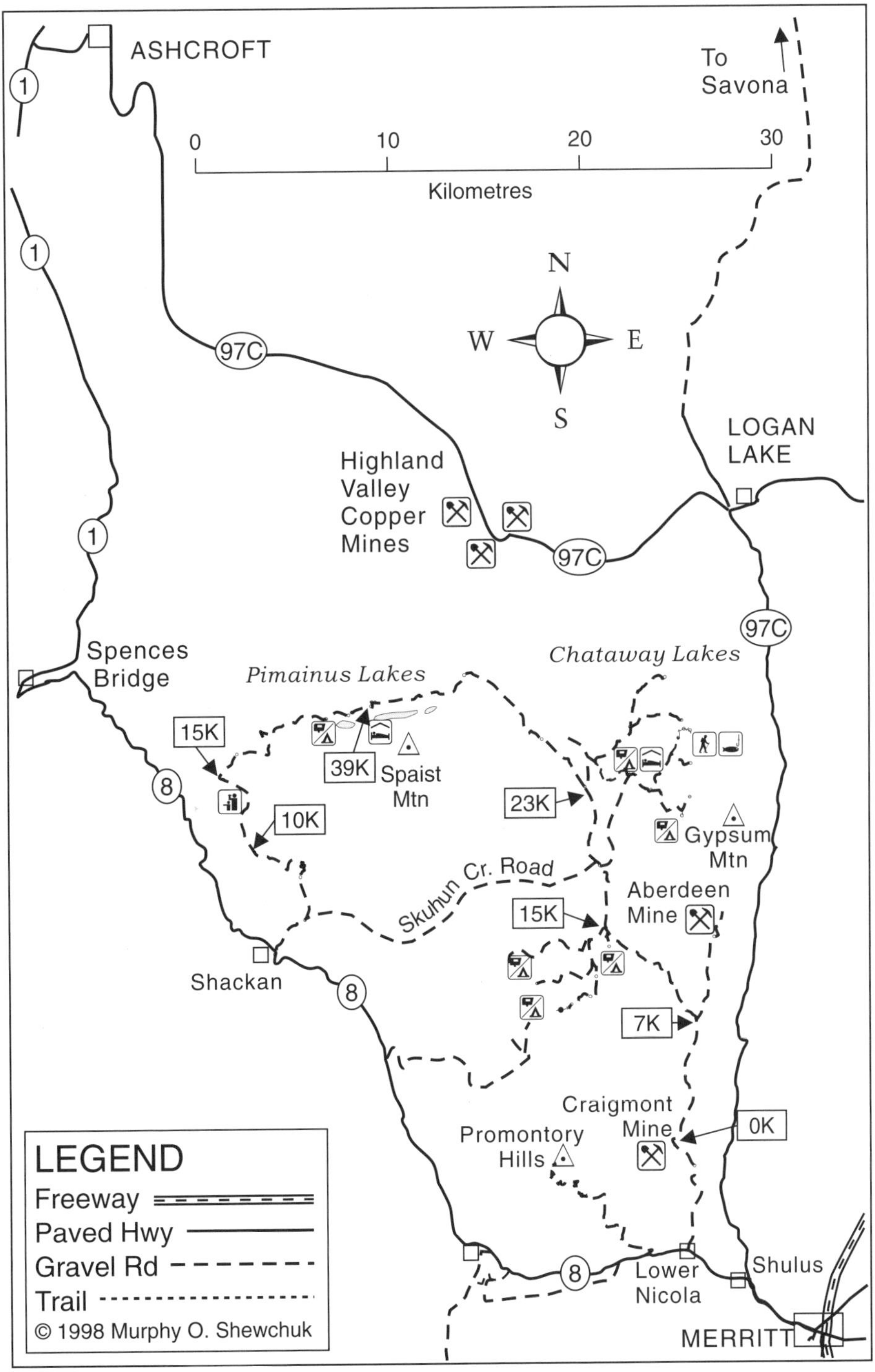

Map 31
Highway 8 and 97C : Merritt - Spences Bridge - Ashcroft - Logan Lake.

23K marker. A right turn here will take you into the northern portion of the Chataway Lakes area. Rough four-wheel-drive roads lead to Roscoe, Billy and Antler lakes.

The Pimainus Lakes signs make the rest of the loop fairly easy to follow as the road swings west. Keep right at the junction with Skuhost Creek Road near the 27K marker, and if you ignore the Skuhun-Spaist Forest Service Road, you should reach Pimainus Lakes Fishing Camp near the 39K marker.

According to Bob Kirkpatrick, the first cabins were built near the lake in the early 1950s. Bob's family has had the fishing camp since the early 1960s and he has been operating it since 1964. The Kirkpatrick's encourage families to use their facilities with special rates for children under 12. Fishing opportunities are plentiful. In addition to the four lakes that make up the Pimainus chain, Calling Lake and Island Lake offer trophy fishing for trout to three kilograms. While dry weather access to Pimainus is easy from either direction, Bob Kirkpatrick describes the six to eight kilometre road to Calling Lake as a "boulder field" that will take a pickup with good clearance upwards of 45 minutes to traverse.

There is also plenty to do for the non-fisher from this base at 1525 metres above sea level. There are hiking trails around the lakes suitable for youngsters and little more strenuous walk to the 1850 metre summit of Spaist Mountain.

The Forest Service Recreation Site is one lake (2.6 km) west of the Pimainus fishing camp. Beyond the recreation site, the roadside markers begin a count-down to Shackan. "Down" is the operative word in the 1100 metre descent to Highway 8. Some sections of the road, particularly between the 10K and 5K markers, are steep (up to 20%). However, some flat sections near the 15K marker offer a good view of the Nicola Valley, the Cascades and the Coast Mountains.

The Pimainus loop joins the previously-mentioned Skuhun Creek Road approximately 21.5 km from the Recreation Site and then reaches Highway 8 about three kilometres later. The distance from Lower Nicola to Shackan via Chataway and Pimainus is about 71 kilometres while the direct route along Highway 8 is about 34 kilometres (21 miles). Half the distance, but not nearly as much fun.

•••

Additional Information:
Chataway Lakes Resort, P.O. Box 294, Merritt, BC V1K 1B8
Kinnikinic Homestead, Box 707, Merritt, BC V1K 1B8
Pimainus Lakes Resort, 5368 - 198 Street, Langley, BC V3A 1E9

•••

43

Chataway Trails

Statistics	For maps, see pages 179, 182 & 184.
Topo Maps:	Mamit Lake 92 I/7 (1:50,000).
Forest Maps:	Merritt Forest District Recreation Map.
Communities:	Merritt & Spences Bridge.

The plateau surrounding Chataway Lake is honeycombed with roads and trails, some originating in the late 1800s when Skuhun Creek was on the route to the Cariboo. Others were used by the first prospectors hauling copper ore out of the Highland Valley. While most of the old pack trails have disappeared, some have been rejuvenated and new trails have been cut to points of interest on the plateau.

Knight Lake Trail

Knight (Echo) Lake lies to the northwest of Chataway Lake and a short walk to the west of Roscoe Lake. Roscoe Lake, at 1585 metres, is one of the highest fishing lakes in the region. It is also shallow in places, with a boulder-strewn bottom. The Forest Service recreation site at the south end of Roscoe Lake is six kilometres northwest of Chataway Lakes Resort over a rough gravel road. You can also reach Roscoe Lake by taking the road to the right (east) near 23.5K on Skuhun-Pimainus Forest Service Road. Continue east for 1.6 kilometres and then take the road north for 3.2 kilometres, keeping left at the major junctions.

A trail from the recreation site winds a third of the way up the west shore of Roscoe Lake before heading west into Knight Lake. (See the map on page 182.) There is a walk-in campsite along the trail, a few hundred metres from the vehicle site. It's less than a half hour walk to Knight Lake, and the trout fishing can be well worth the effort, even when the rest of the lakes in the area are slow. If you are interested in an overview of the plateau, hike the short side trail to the top of Roscoe Bluff.

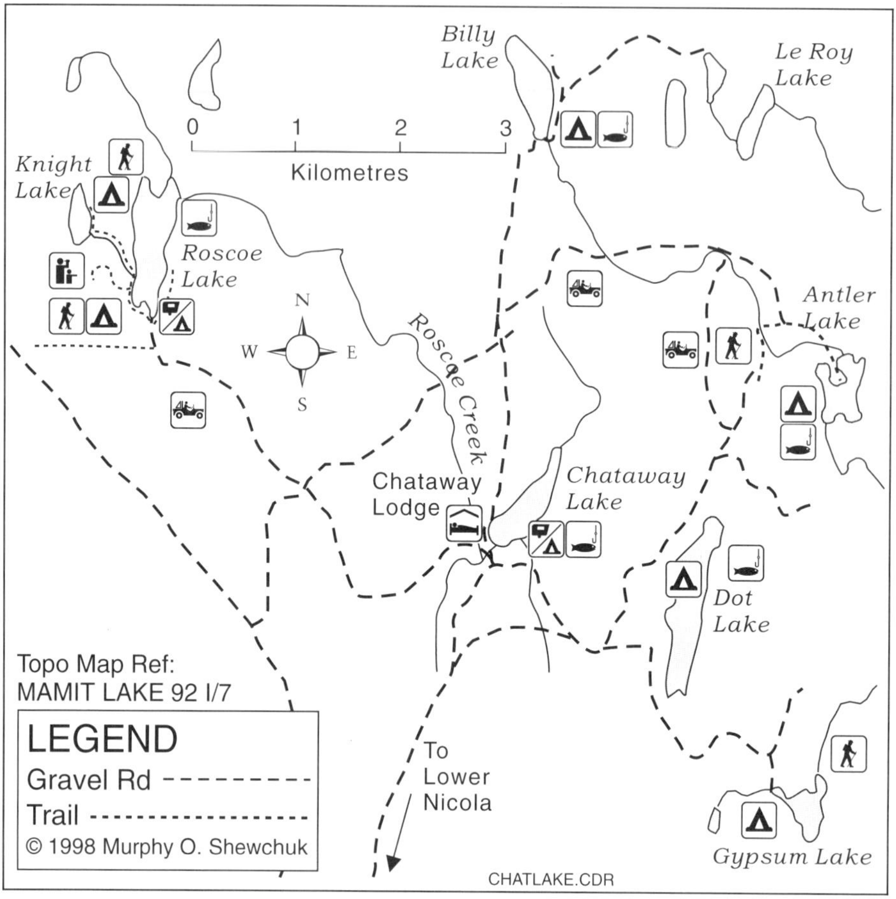

Map 32
Chataway Lakes area.

Antler Lake Trail

Antler Lake, northeast of Chataway Lake, is accessible by road and trail from Dot Lake or via Billy Lake Road. While you can drive in to the Forest Service recreation site at Antler Lake via the Billy Lake road, the route from Dot Lake, about three kilometres southeast of Chataway Lake, presents an interesting hike through the pines.

To get to the trailhead, follow the Dot Lake road southeast of the Chataway Resort gate. With the gate as km 0, keep left at km 0.3 and again at km 1.7 and 2.2. You should reach the Dot Lake recreation site at km 3.3. Go left (north) at the recreation site for 2.1 kilometres, ignoring the side roads to the right and left. Watch for markers on the right (east) at a wide spot in the road, indicating the start of the Antler Lake trail.

Figure 43
A fisherman tries his luck on Antler Lake.

From the trail head to the recreation site is a well marked 20-minute (0.8 km) hike through the open timber. It is another 0.4 to 0.5 kilometres to the lake on a trail that climbs over a knoll and meanders through a marsh on a boardwalk. (Kids will love it!) On one of our visits, a pair of float-tube fly fishermen were enjoying the warm afternoon sun while drifting lazily on one of the lake's many finger-like bays.

Gypsum Mountain Trail

Gypsum Mountain,(for maps, see pages 182 and 184) east of Gypsum Lake and almost due south of Antler Lake, is the site of an interesting 1.5 hour (2.0 kilometre) hiking trail. Access from Chataway Lakes Resort to the Gypsum Lake trailhead is the same as to Dot Lake for the first part. With the resort gate as km 0, keep left at km 0.3 and again at km 1.7. Go right at km 2.2 and continue southeast past Dot Lake at km 2.9. Keep left at the junction near the 27K marker (km 4.0) and go right on a rough road near km 5.0. It's only 0.3 kilometres to the Gypsum Lake trailhead, so if you're not impressed with the state of the road, you could hike to the recreation site and the lake.

The start of Gypsum Mountain trail is well marked. (See the map on page 184.) It skirts the north shore of the lake for about 0.5 kilometres to a rocky point. It then heads south for about 0.4 kilometres along what appears to be an old mineral survey line. It then crosses a dam and begins a gradual climb east along another survey line for about 0.6 kilometres before beginning a steeper climb to a saddle between the twin summits.

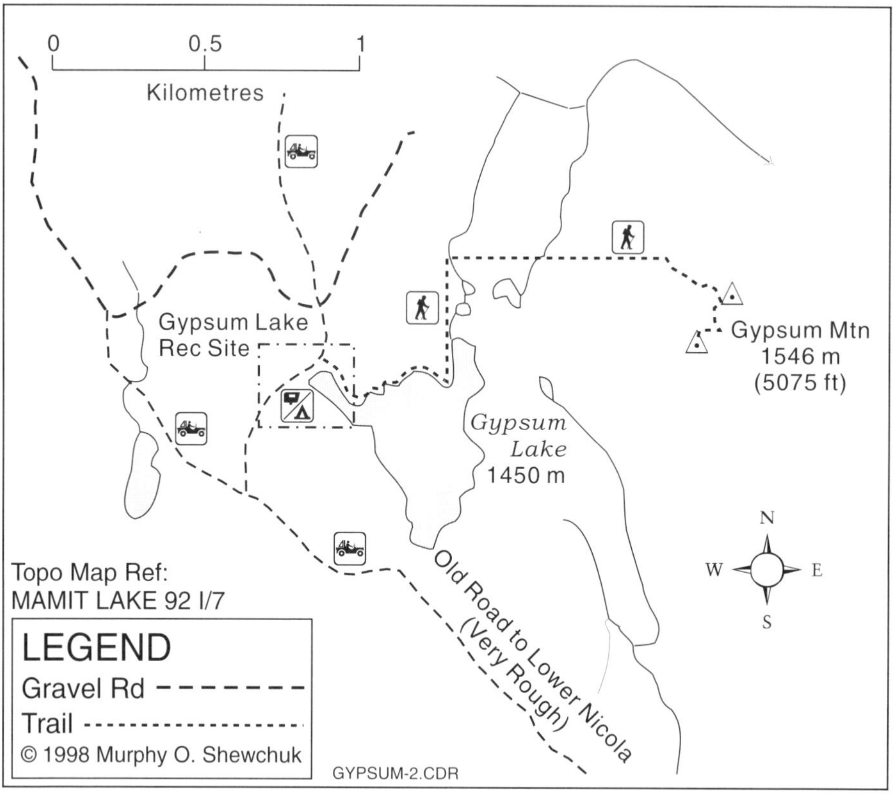

Map 33
Gypsum Lake and Gypsum Mountain Trail.

From the saddle, its a short hike to either open peak. When we last visited the mountain, there was a pit toilet at the saddle and a few choice spots to pitch a tent — but no drinking water.

•••

GPS References for major points of interest. See page 9 for details.

Wpt	Description	Latitude	Longitude
W01	Gypsum Lake Rec. Site	N50° 21' 01.056"	W120° 52' 28.392"
W02	Gypsum Mountain	N50° 21' 05.304"	W120° 51' 20.550"
W03	Roscoe Lake Rec. Site	N50° 23' 25.218"	W120° 57' 30.084"
W04	Roscoe Bluff	N50° 23' 37.404"	W120° 57' 48.138"
W05	Knight Lake	N50° 23' 49.200"	W120° 57' 59.898"
W06	Leroy Lake Rec. Site	N50° 24' 24.468"	W120° 54' 16.710"
W07	Billy Lake Rec. Site	N50° 24' 48.450"	W120° 53' 02.436"
W08	Antler Lake Rec. Site	N50° 23' 21.798"	W120° 52' 16.014"
W09	Chataway Resort	N50° 22' 17.046"	W120° 54' 54.216"
W10	Dot Lake Rec. Site	N50° 22' 09.324"	W120° 53' 24.432"

•••

44

Promotory Mountain Road

Statistics	For maps, see pages 179 and 186.
Distance:	13.5 kilometres, Highway 8 to Promontory Lookout.
Travel Time:	One half hour.
Condition:	Gravel, rough in places.
Season:	June through October.
Topo Maps:	Merritt 92 I/2 (1:50,000).
Forest Maps:	Merritt Forest District Recreation Map.
Communities:	Merritt and Lower Nicola.

The Forest Service lookout on 1734 metre high Promontory Mountain offers one of the best views in the area. The Cascades, Coast Mountains, Cariboo Mountains and Monashee Mountains are all visible from the catwalk.

The 13.5 kilometre road to Promontory Mountain leaves Highway 8, 11 kilometres northwest of Merritt.

Follow Woodward Road to the refuse dump, then turn left at the gate. The rough road climbs through an old burn and then through stands of pine, fir and aspen. Side roads lead to hayfields and old logging clearings, but the major junctions are marked with simple "L/O" signs. The road is seldom open to the top before mid-May. In dry weather it is passable by car or light truck with ample clearance.

•••

Figure 44
Promontory Lookout.
(Photo 234-E-27A)

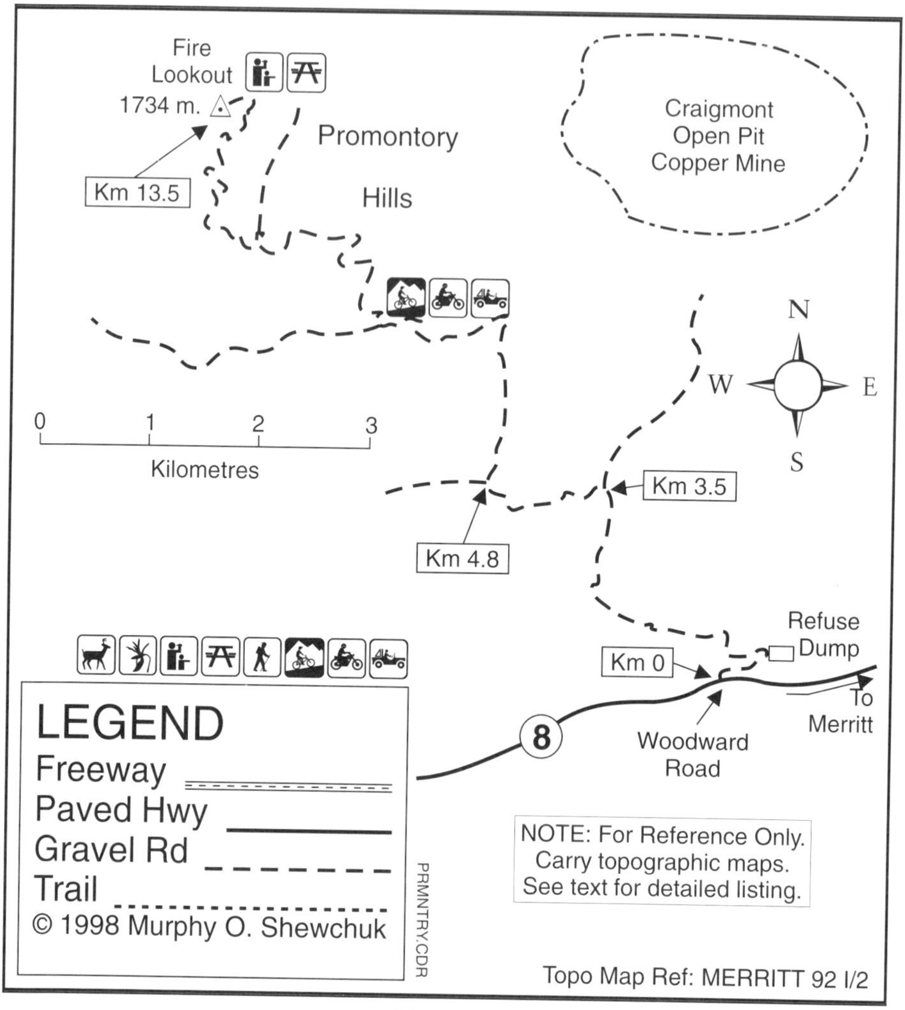

Map 34
Promontory Mountain

GPS References for major points of interest. See page 9 for details.

Wpt	Description	Latitude	Longitude
W01	Jct. Hwy 5 and Woodward Road	N50° 08' 59.832"	W120° 54' 46.356"
W02	Nicola Valley viewpoint	N50° 09' 16.074"	W120° 55' 25.716"
W03	Jct. and cattleguard go left	N50° 09' 54.810"	W120° 55' 43.770"
W04	Jct. go right	N50° 09' 55.626"	W120° 56' 34.698"
W05	Jct. crossroads go straight	N50° 10' 32.856"	W120° 56' 29.250"
W06	Promontory Lookout	N50° 11' 42.456"	W120° 58' 19.812"

•••

45

Stoyoma Mountain Trails

Statistics **For maps, see pages 188, 190 and 191 .**

Distance:	40 kilometres, Highway 8 to Cabin Lake.
Travel Time:	Two to three hours.
Elevation gain:	1150 metres.
Condition:	Gravel, rough sections on Stoyoma Mtn.
Season:	June through September.
Topo Maps:	Boston Bar 92H/14 (1:50,000). Prospect Creek 92I/3 (1:50,000).
Forest Maps:	Merritt Forest District Recreation Map.
Communities:	Merritt & Lower Nicola.

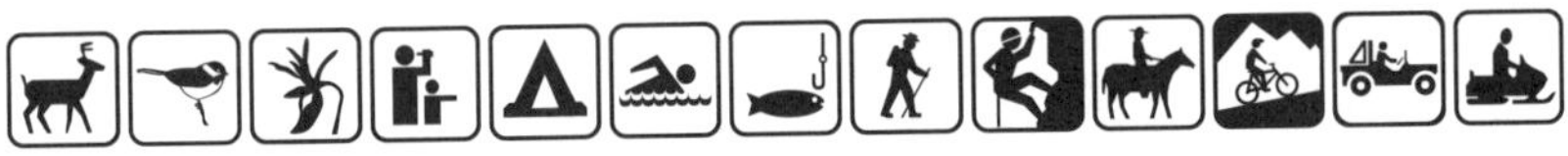

Stoyoma Mountain, at 2282 metres, is one of the highest peaks on the north slope of the Cascades. Half a dozen lakes are nestled in the folds of Stoyoma's slopes, and a maze of sheep trails wind through the alpine meadows and sub-alpine timber. Stoyoma is also home to one of the few resident mountain goat populations in the region.

An access road was cut to Cabin Lake many years ago and is still passable on foot or horseback — or with a four-wheel-drive vehicle with good clearance. Weyerhaeuser Canada Ltd.'s logging roads provide easy access to the 1700 metre level for the weekend traveller. Snow and washouts may keep the upper reaches of the road closed until July and early snowstorms can make it dangerous again in October.

To find Stoyoma Mountain, drive 20 kilometres west of Merritt on Highway 8, then follow the Petit Creek Road (which soon becomes Prospect Creek Forest Service Road) south and up Spius Creek. Keep right at the junction near km 3 — the road to the left returns to Highway 8 as the west end of Sunshine Valley Road.

Spius Creek has cut a fairly narrow canyon throughout its length and Prospect Creek Forest Service Road stays above the canyon, passing cattle and sheep ranches along the way. There are two side roads at a junction near the Prospect Creek bridge at km 24. Keep to the left at both junctions, and begin the steep climb out of the valley. The road levels off somewhat near km 31, having climbed over 500 metres from the Prospect Creek bridge. It crosses another clear, cold creek near km 34.5 — an excellent spot to fill your water bottles.

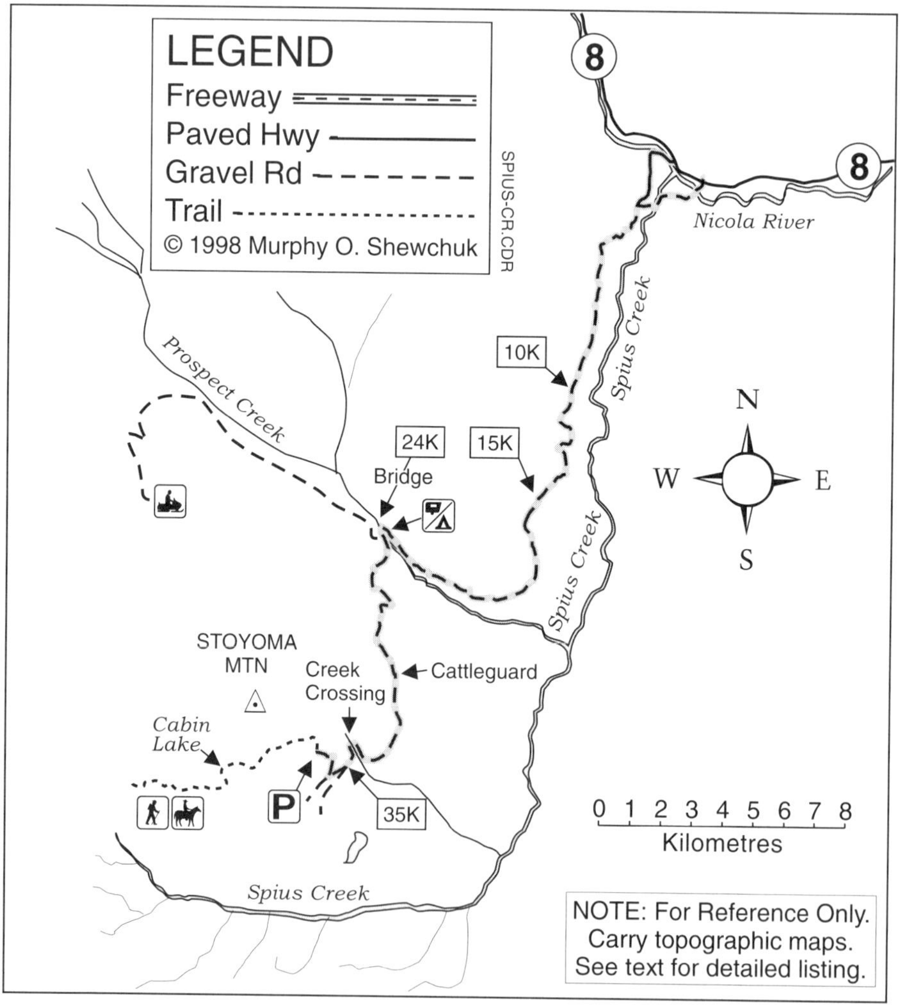

Map 35
Petit Creek Road / Prospect Creek Forest Road to Stoyoma Mountain area.

The kilometre signs along the route serve as a guide. Just beyond the 35K sign (tacked to a tree or stump well above the road), a side road climbs to the right. Take it and watch for a junction 0.8 kilometres farther along. The road straight ahead continues into logging areas on the lower slopes of the mountain. Nestled in a hollow five minutes walk from a log landing a kilometre farther along this road is beautiful Lightning Lake. The trail into Lightning Lake is steep, but has been driven with a four-wheel-drive vehicle.

Figure 45
The view west from 2282-metre-high Stoyoma Mountain. (Photo 272-A-4)

If Cabin Lake or Stoyoma Mountain is your destination, take a right turn at the junction 0.8 kilometres from the main road and follow this road through a logged strip and into a timber buffer zone, keeping left at a junction at 1.3 kilometres. Just as the road emerges from the buffer strip, (km 2.3) an older road skirts the edge of the uppermost logged area and climbs through the trees to the west.

This is the original cattle and sheep trail into Cabin Lake. For most two-wheel-drive vehicles, this is a good place to park and start walking. In a 4x4 with good clearance, you may be able to continue the remaining 3.5 kilometres to Cabin Lake, but turnaround space is limited.

Cabin Lake takes its name from an ancient log cabin that sits on a point that juts into the lake. Despite its elevation of 1860 metres, Cabin Lake can occasionally be warm enough for swimming in mid-summer. A Forest Service recreation site at the lake can be a good base for a Stoyoma Mountain ramble. Come equipped with good boots, topographic maps and plenty of time.

One trail worth exploring leads into a basin where a water bomber crashed a number of years ago, to the southwest of Cabin Lake. The trail begins near the outlet of the lake and, after a short, steep climb, follows

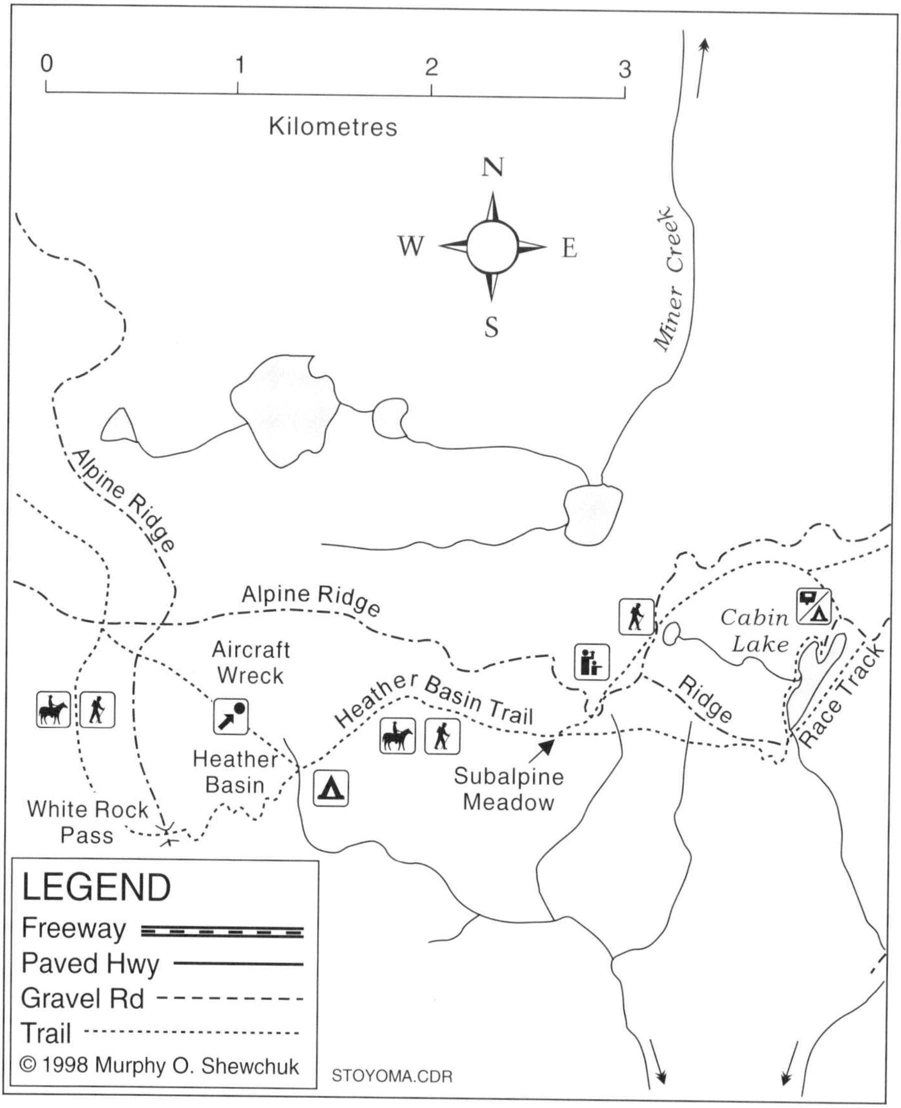

Map 36
Stoyoma Mountain area — west section.

along the mountain slopes through the timber. It ends at the base of several rocky slopes. If you look up to the north, you may be able to see the wreckage part way up the mountain side.

On your way back, you can climb up and over the crest of the ridge through one of the meadows nearer the lake and ramble back down along the creeks that feed into Cabin Lake. Watch carefully, and you may find one of the game trails that lead downhill around the rock outcroppings and through the timber. Look southwest from the top of the ridge on a

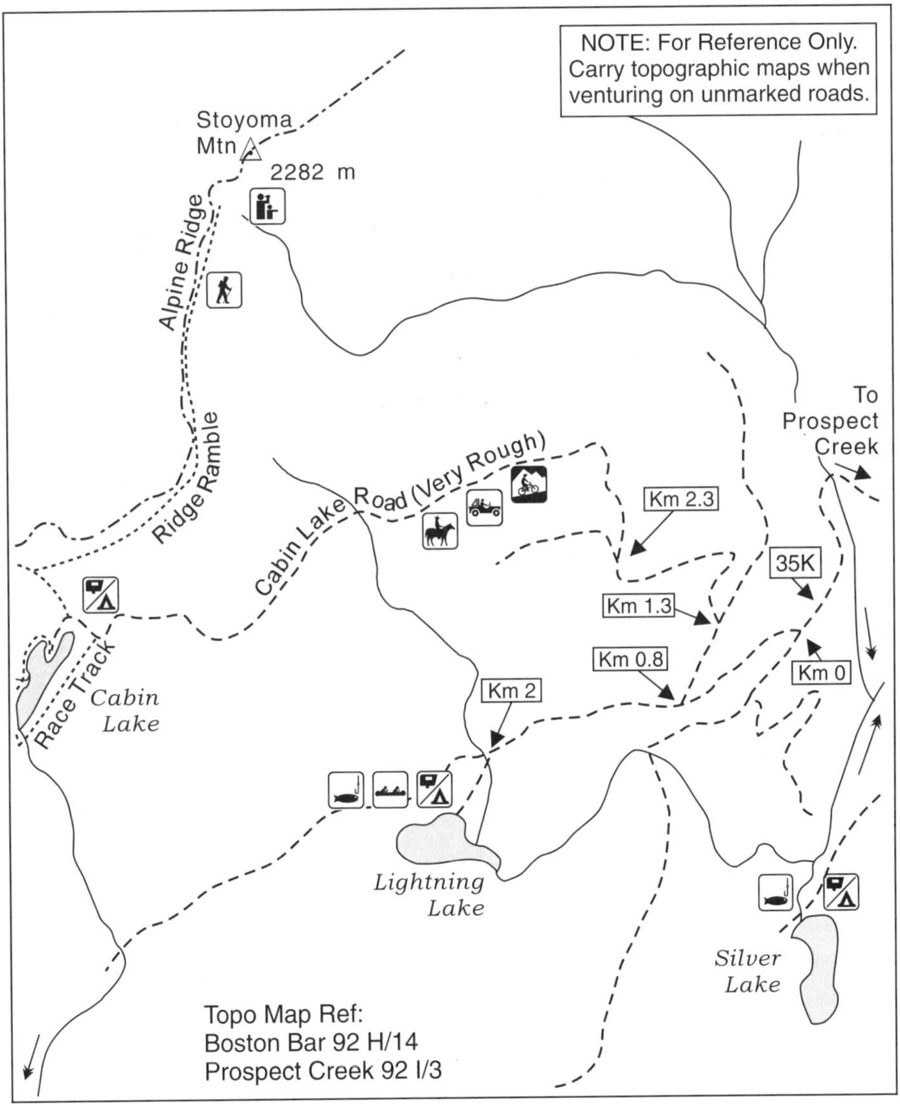

Map 37
Stoyoma Mountain area — east section.

clear day and you may be able to see the snow-capped peak of Mount Baker on the distant horizon.

Stoyoma Ridge is also easily reached from Cabin Lake or from almost any meadow along the last stretch of road into the lake. The subalpine timber is sparse and provides little opposition to the determined explorer. Although not particularly difficult, the round trip from 1860 metre Cabin Lake to the 2282 metre peak will take the better part of a day. Carry drinking water, as safe water may be in short supply along the ridge.

Figure 46
Nola, Bridget and Katharine on the ridge south of Cabin Lake. (Photo STO-112)

•••

GPS References for major points of interest. See page 9 for details.

Wpt	Description	Latitude	Longitude
W01	Highway 8 & Petit Creek Road	N50° 08' 49.560"	W121° 01' 06.220"
W02	Hwy 8 & Sunshine Vly Road	N50° 08' 34.260"	W121° 00' 23.671"
W03	Prospect Creek FS Road	N50° 08' 08.820"	W121° 02' 12.641"
W04	Davidson Road	N50° 07' 37.203"	W121° 02' 48.338"
W05	24K marker and bridge	N50° 02' 35.408"	W121° 09' 01.823"
W06	Cattleguard and gate	N50° 00' 06.934"	W121° 08' 37.129"
W07	Creek crossing	N49° 58' 56.013"	W121° 09' 50.064"
W08	Turn right off main road	N49° 58' 34.025"	W121° 09' 54.065"
W09	End of two-wheel-drive road	N49° 58' 46.512"	W121° 10' 46.717"
W10	Route to Peak	N49° 58' 51.386"	W121° 11' 54.175"
W11	Cabin Lake	N49° 58' 31.073"	W121° 13' 15.322"
W12	Switchback climb	N49° 58' 09.111"	W121° 13' 29.687"
W13	Top of rocky slope	N49° 58' 09.083"	W121° 13' 38.423"
W14	Meadow route over ridge	N49° 58' 12.708"	W121° 14' 10.120"
W15	Heather Basin & crash site	N49° 58' 12.257"	W121° 15' 44.723"

•••

46

The Land

Physical Features

The Coquihalla River begins northeast of Hope at the summit of the Coquihalla Pass. Its source, the Coquihalla Lakes, is surrounded by craggy mountain peaks, some reaching 2130 metres in elevation. As the river flows southward, it is joined by numerous small creeks and a few more significant waterways. Boston Bar Creek, the first major tributary, has its source in a mountain valley a short distance from the headwaters of the Coquihalla. It, however, flows to the west of Needle Peak, joining the Coquihalla downstream from the summit.

The Nicola River rises in the upland country of the Douglas Plateau, part of the much larger Thompson Plateau that separates the Coast Mountains to the west from the Monashee Mountains to the east. In its 150-kilometre journey from its origins near Pennask Lake to the Thompson River at Spences Bridge, the Nicola has etched a path through a variety of climatic and topographic zones.

The pine-forested lake country where the river originates lies as much as 2,000 metres above sea level. As it winds southwestward, the Nicola River passes through an ever-changing panorama of rolling natural bunchgrass. This is Douglas Lake Ranch country, famous for its Hereford cattle. The river soon widens to form Douglas Lake and then Nicola Lake. Two other important creeks drain the plateau country, emptying into Nicola Lake: Stumplake Creek empties the string of lakes to the north, and Quilchena Creek carries the water off the lake-studded plateau country to the southeast.

The Nicola River, lined with cottonwoods, cuts a meandering path through the hay fields on the way from Nicola Lake to the junction with the Coldwater River in the heart of Merritt. The Nicola's major tributary, the Coldwater rises in the North Cascade Mountains, west of the Coquihalla Lakes. Spire-like snow-capped mountain peaks, glacial silt-laden streams and steep timbered slopes present a sharp contrast to the plateau country. The natural grasslands that are so much a part of the Nicola Basin are not apparent in the Coldwater Valley. Instead, somewhat heavier rainfall and differing soil conditions have fostered heavier timber growth in the upper reaches of the U-shaped valley; nearer Merritt, scattered ponderosa pine is found.

From Merritt, the Nicola River strikes out in a northwesterly direction, following an ever-narrowing valley through the north slope of the Cascades. At first the valley floor is carpeted with small ranch holdings, irrigation sprinklers creating man-made rainbows over the alfalfa fields. But as the river nears the Thompson, the fields give way to steep, colorful canyon walls. Striking hoodoos stand guard over the tiny market gardens while outside the reach of cultivation, sagebrush and prickly pear cactus thrive.

Figure 47
Jerry McKenzie, near the Douglas Lake Road. (Photo DLG-360)

If you drive slowly up the Douglas Lake Road from the junction with Highway 5 at Nicola Lake just as the morning sun lights the hills you will see a segment of the physical history of the valley that dates back more than 100 centuries. Across the valley to the north, the morning shadows accentuate the shorelines of two ancient lakes that once filled the valley from ridge to ridge. The lower shoreline is visible behind the red-painted buildings of the Quilchena Cattle Company "Home Ranch" while the upper shoreline can be seen as you approach BC Hydro's Nicola substation.

During the last period of glaciation, which ended some 10,000 years ago, almost all British Columbia was covered by massive sheets of ice, often reaching thickness in excess of 1,500 metres. Only the highest peaks escaped the grinding action of this flowing mass of destruction. The Thompson Plateau region was ripped open and reformed, leaving scars that the dry climate and open grasslands have preserved.

The many north-south lying chains of lakes indicate the direction of the ice movement. Drumlins and eskers, side and front moraines, and erratics such as the huge boulders in the open fields near the north end of Stump Lake, mark where the ice melted, depositing its heavy load.

The first major lake to form in the Nicola Basin as the glaciers began to melt was Glacial Lake Quilchena. It reached an elevation of approximately 1000 metres above sea level and drained southward into the headwaters of Otter Creek near Aspen Grove. Shore features of this ancient lake are only visible in a few locations.

Glacial Lake Hamilton was formed when the ice plug retreated in the Salmon River Valley permitting eastward drainage through the Douglas Plateau. This lake was only about 70 metres lower than Glacial Lake Quilchena, but some of its features are still visible in the form of the previously mentioned shoreline, and terraces on the north slope of Mount Hamilton, southeast of Nicola Lake.

The last of the giant glacial lakes was Lake Merritt. It was formed when the ice melted out of the Campbell Creek Valley to the north and allowed the meltwater to drain into the Thompson Valley east of Kamloops. The lake level was restricted by a hard layer of bedrock located between Stump and Napier lakes, alongside Highway 5A. A significant canyon was carved into the reddish rock as Glacial Lake Merritt spilled over this outcropping. As the warming climate forced the giant ice sheet northward, the last ice plug melted out of the northwestern (lower) reaches of the Nicola River, opening the present drainage route out of the valley.

Beneath the grasslands, lakes and forests of the Thompson Plateau and the North Cascade Mountains lies a stone heart of great richness and variety. The plateau country contains sedimentary and volcanic formations into which have intruded stocks of granitic rock. Flat or gently sloping ancient lava beds obscure even older rock formations. These lava beds often underlie the step-like slopes and the large unbroken plateau areas. Coal deposits and fossil beds in the immediate Merritt area are evidence of the age of some of the rock formations. As will be discussed in the section on settlement, these coal deposits once played a significant part in the development of the valley.

The Serpentine Belt which crosses the North Cascades from the vicinity of Ladner Creek in Coquihalla Canyon to just north of Boston Bar on the Fraser, is a mineral-rich zone. Carolin Mines, which has workings near Ladner Creek, was considered to have the potential of a large tonnage gold mine. Other known minerals in this region include silver, lead, iron, zinc, copper and molybdenum. Rockhounds and artists find the Serpentine Belt of interest because of the nephrite jade and soapstone deposits that are found in conjunction with serpentine.

Figure 48
An open-pit copper mine in the Highland Valley. (Photo LOL-019)

The most significant geological feature of the Nicola Valley is the Guichon Creek Batholith. This 200-million-year-old intrusion of plutonic rock extends from Merritt to Cache Creek and contains some of British Columbia's largest deposits of low grade copper and molybdenum. The former Craigmont Mine near Merritt and the world-class open pit mines of the Highland Valley are situated on the batholith. (See *Merritt to Ashcroft*, page 131.)

Climate and Vegetation

Cactus and columbine, sagebrush and cedar. These may be unlikely plant combinations, but they are all found in Coquihalla Country, often within hiking distance of each other. The Nicola drainage basin, with its widely varied terrain and elevation range, lends itself to an extended mixture of climate zones and, consequently, a diverse flora.

In general, the Thompson Plateau has a continental climate. The winters are moderately cold and the summers hot. Winter temperatures at Merritt, for instance, range from -30 degrees Celsius to 0 degrees Celsius and temperatures during July and August easily reach 30 degrees Celsius and occasionally may go as high as 40 degrees Celsius. Annual precipitation varies from 15 to 75 centimetres; the average at Merritt is 30 centimetres. Temperature and precipitation variations are closely related to elevation.

Snowfall is usually light in the main valleys with most of the precipitation falling in the form of rain in the spring months. The vegetation at the lower altitudes is typical of a hot, dry climate with sagebrush, bunchgrass, rabbit bush and ponderosa pine being the natural dominant species. A close look at the ground in the early spring may reveal sage buttercups, yellow bells, shooting stars and a variety of other tiny wildflowers. In late April and early May, the sunflower-like blossoms of arrowleaf balsamroot turn the otherwise drab hillsides into a field of yellow. They are soon joined by the white blossoms of saskatoon bushes, fields of blue lupines and then a colour kaleidoscope that includes yellow arnica, red and pink wild roses, rose-purple sticky geranium, lavender blue penstemon and other flowers including the pale yellow prickly pear cactus.

As the elevation and rainfall increases, the ponderosa pine give way to stands of interior Douglas fir. Trembling aspen and cottonwood thrive along the creek beds and in the wetter hollows. The flowering plants change, too, though the transition may not be as distinct. Wild strawberry, Oregon grape, stonecrop, fireweed, western columbine and Columbian lily are but a few that add their beauty to the landscape in spring and summer.

Snowfall is usually heavier at the higher elevations of the plateau country. As much as one metre of snow can accumulate over the winter, but warm chinook winds often strip bare the lowland valleys and exposed slopes.

The Coquihalla Valley, which is part of the freeway route linking Hope and Merritt through the North Cascades, is an area of heavy snowfall. Records kept when the Kettle Valley Railway was operating (1916-1963) indicate that snow accumulations of over 12 metres were not uncommon and, in years of heavy fall, snow depths reached 15 metres. The most pronounced climatic transition in the region takes place in the southern portion of the Coquihalla Valley where average annual precipitation changes from 100 to 40 centimetres in less that 24 kilometres.

The sub-alpine zone that makes up much of the North Cascades and portions of the Thompson Plateau is covered by sub-alpine fir and Engelmann spruce. Shrubs and berry bushes are plentiful, including tasty huckleberries and wild strawberries. Many of the wildflowers that grow on the lower slopes can be found in even greater profusion in logged areas, old burns and snowslide paths. Indian paintbrush in a variety of colors, yellow and purple violets, wine-red monkey flower and yellow avalanche lilies all add their beauty to the spring and summer scene.

The alpine tundra that makes up a very small portion of the region is not as barren as it first appears. Many varieties of trees and flowering plants have adapted to the harsh winters and short summers. Some are adaptations of plants found at lower elevations while others are unique to the alpine environment. One of the more distinctive flowering plants found here is the western anemone. It forces its shoots up through the edges of the melting snow, quickly opens its leaves and then its blossoms to catch the warm rays of the sun. As the flower stem grows, its "towhead baby" seed pod is thrust into the wind where the parachute-like seeds are widely dispersed.

Ranching is the predominant form of agriculture in the Nicola Valley. Hay crops grow on the bottom land while cattle graze on the upland grass, sometimes into the alpine. During the winter, the cattle are driven down to the lowlands and fed hay produced during the previous summer. Irrigation is used to overcome the low moisture holding capacity of the sandy soils and to augment the limited summer rainfall. Thus man has modified the effects of climate and terrain in his attempt to wrest a living from the land.

•••

47

Wildlife

Birds

The relatively mild climate and abundance of small lakes makes Coquihalla Country a paradise for birds. Estimates vary, but knowledgeable naturalists suggest that between 125 and 200 species of birds are at least seasonally present in the Thompson Plateau region.

Mid-winter bird counts carried out by members of the Kamloops Naturalist Club over the past decade have identified at least 90 species. The total number of birds counted in the specified four-hour count period have exceeded 3,000. Among the most common on these winter counts are Bohemian waxwings, starlings, house sparrows, snow buntings, mallard ducks, Canada geese and whistling swans. By far the predominant species are the whistling swans which winter on Shuswap Lake, the South Thompson River and, in mild winters, on Nicola Lake. These beautiful white birds have a wing length of up to 560 milimetres and weigh as much as 10 kilograms, placing them among the largest flying birds. They are likely to be mistaken only for their larger cousin, the trumpeter swan.

On one January count the Kamloops Naturalist Club reported 565 swans with at least 156 immatures. Twenty-four whistling swans, including eight immatures, were counted at the mouth of Quilchena Creek — the only open water on Nicola Lake at the time.

It would be impossible to do the Nicola Valley bird population justice in the limited space available in this book. However, the following season-to-season review will highlight some of those that I have observed.

Winter sightings include the whistling swans, great gray owls, the occasional snowy owl, various birds that frequent backyard feeders and a stray blue heron that wanders the shores of the Nicola River. Spring brings a large number of bald and golden eagles, plus the occasional osprey. Canada geese and waterfowl are plentiful, particularly on the smaller lakes and ponds of the plateau country. The bright flash of the mountain bluebird is also a frequent sight, and nesting boxes have been set up on the grasslands east of Stump Lake to encourage these insect-eaters to stay. Chukkar partridge and several species of grouse are common throughout the region. They can most often be seen in the scattered timber in the summer and early autumn. In the autumn, the region becomes part of the migratory waterfowl flyway and ducks, geese and bobbing rafts of coots are a familiar sight on Nicola and Stump lakes.

Figure 49
A black bear next to a rose bush. (Photo 429-C-16)

Mammals

Few of the larger mammals of the Thompson Plateau and North Cascade Mountains live without some form of contact with man. Hunting is allowed in season and poaching is sometimes a problem. Because there are no large federal or provincial parks to offer sanctuary, the mammals tend to remain wary of man and his presence.

Despite this justified wariness, the most visible of the large mammals is the shy mule deer. This adaptable animal can be found throughout the region, from valley floor to alpine meadow. Mule deer are often seen in the early morning hours, drinking from a stream or feeding on the tender shoots of young aspen or cottonwood. Whitetail deer inhabit only a limited portion of the valley. In the summer months, their main habitat is the slopes of Promontory Mountain and the lower Guichon Creek valley northwest of Merritt. A small herd of whitetail deer is also known to frequent the Douglas Plateau to the north of Chapperon Lake.

Moose, although not originally native to the southern plateau country, migrated southward from central British Columbia as man opened up the heavy forest barrier of the north Cariboo with his logging, farming and roads. Moose are now dispersed throughout the plateau and are as likely to be seen in the open grasslands as in the upland swamps.

Mountain goats range the skyline ridges of the North Cascade Mountains. Stoyoma Mountain and the peaks of the upper Coldwater Valley, to the west of the Coquihalla Lakes, are the best places to catch a glimpse of these white ungulates.

Black and brown bear are relatively common in the Nicola basin, although they are generally wary of man and will usually run at the first sign of a confrontation. ("Usually" is the key word. Almost every animal will fight if it fears that it or its young are threatened.) Though seldom seen, there are still a few grizzly bears in the North Cascade Mountains.

Coyotes are the grassland scavengers. They feed on small animals and birds and sick calves or dead cattle. They are the most elusive of the grassland mammals: their bounding "bush" is often all that is seen as they dodge among the sagebrush. It is the lucky photographer who manages to "shoot" a coyote in the wild.

It is an even luckier photographer who manages to capture a mountain beaver on film. Not a true beaver, but a large primitive rodent up to 40 centimetres in length and weighing more than one kilogram, this nocturnal mammal lives near the slow moving watercourses of the Cascade rain forests. The mountain beaver habitat can be recognized by the neat stacks of bracken and other lush vegetation found drying outside their tunnel entrances. Though thought to be restricted to a small strip along the west coast of North America, sightings have been reported in both the Nicola and Coquihalla watersheds.

In all, some 50 species of mammals have been recorded in the region. In addition to those already named, others occasionally seen include cougar, lynx, beaver, weasel, mink, fox and, in the alpine, marmot, pika and bog lemmings.

Insects

The avid outdoor enthusiast has found Coquihalla Country refreshingly free of bothersome insects. Mosquitoes can be somewhat of a nuisance early in the season as can yellow-jacket wasps in late summer. Shy black widow spiders inhabit the darkness of some abandoned barns and cabins, but their bite is not considered dangerous at this northern latitude.

The hot dry summers and generally open terrain do not encourage mosquitoes, blackflies or horseflies. Most summer hiking and fishing trips can be accomplished with minimal use of insect repellent or special clothing, although both are good insurance.

One of the most feared and least understood creatures of the Nicola Valley is the Rocky Mountain or paralysis tick. Also known as the wood tick, it ranges in Canada from the British Columbia interior dry belt to Saskatchewan. In the Nicola Valley, the wood tick is misnamed, for it inhabits the open grasslands, preferring rocky southern exposures where the sun warms the earth early in the spring. From March to June, with peak activity in mid-April, ticks rest on the scrub vegetation and bunch-grass, awaiting the unwary warm-blooded passerby. Tick paralysis is a disease associated with the local tick, but symptoms usually occur only after the adult female Rocky Mountain tick has been feeding for about five days. There is no pain or fever as the paralysis starts with the body extremities and progresses towards the core. Complete recovery begins as soon as the tick is removed. If the tick is not removed, death can occur when the vital organs become paralyzed.

Recognition, protection and after-hike hygiene are the basic steps to avoiding tick problems. The adult female Rocky Mountain tick is reddish-brown with a white shield near her head. She is not much larger than a capital "O" on this page. Smooth pants tucked into boots help keep the tick from gaining a firm hold when it is brushed against. A close personal inspection, particularly under wristbands, clothing straps and body hair is necessary to remove any ticks that may have become attached.

If a tick appears to be firmly attached, it can usually be removed with a slow gentle pull. Unlike other varieties with barbed mouthpieces, the Rocky Mountain tick fastens itself to the skin with a whitish glue which usually comes off with the tick.

Amphibians and Reptiles

Coquihalla Country is home to an estimated 14 species of amphibians and reptiles, the most common of which are frogs and garter snakes. There have, however, been sightings of rattlesnakes and bull snakes near Spences Bridge and in the southeast near the Okanagan Valley, though seldom near populated areas. None of the trips outlined in this book is likely to place the hiker in contact with any dangerous snakes.

The bull snake is a shy reptile that would rather retreat than fight. However, when cornered, it will coil, emit a loud warning hiss, and bang its tail against the gravel to imitate a rattlesnake. If forced to strike, it is not poisonous but can inflict a painful bite.

Figure 50
A wary bull snake watches a wary photographer. (Photo 85-A-5A)

The Northern Pacific rattlesnake has many of the same characteristics. Most reported bites have occurred when people inadvertently stepped on or handled a rattlesnake or when the snake was teased. As two-thirds of the bites occur on the feet and lower legs, boots and sturdy pants should be worn. Extra precautions should be taken when climbing to avoid accidentally putting your hand on a sunning snake.

In the event of rattlesnake bite, the important thing is to remain calm and quickly seek medical aid.

Fish

The Thompson-Okanagan Fish and Wildlife management area attracts close to 50 per cent of British Columbia's total freshwater sports fishing activity. When we consider that this area comprises less than 10 per cent of the province, the figure becomes even more significant. As a freshwater sports fishery, the Thompson Plateau is unsurpassed in British Columbia. The Nicola Valley motto of "a lake a day as long as you stay" is no idle boast.

Fish, fresh or dried, were a year-round mainstay of the Indians that inhabited British Columbia prior to the arrival of the first fur traders. Then dried fish and fish oil became important items of barter and were often the winter staple at the Interior fur trading posts. Sockeye salmon were the most valuable fish in the river systems that included the Nass, Skeena and Fraser.

The salmon, and to a lesser extent the smaller species of fish, were thoroughly utilized by the native people. They were eaten fresh, or split and dried for use in winter. The roe was considered a delicacy and stored in special woven baskets for winter use. Fish heads were used as soup stock or in a mixture for tanning hides.

The Coquihalla River and the Coldwater-Nicola system were, and continue to be, important salmon spawning tributaries of the Fraser River. The smaller tributary streams also play host to the eggs of rainbow trout, Dolly Varden char and kokanee. The Coldwater River is a spawning channel for chinook and coho salmon, particularly the stretch of river upstream from Kingsvale.

Chapperon, Douglas and Nicola lakes were considered especially valuable by the Okanagan Indians because of their ample fish stocks and because the ice left them earlier than many other of the plateau lakes. Nicola Lake, noted for its depth, is said to harbour 26 varieties of fish, some weighing up to 9 kilograms: among them are mountain whitefish, spring salmon, kokanee, rainbow trout, Dolly Varden char, Pacific lamprey and freshwater ling cod.

Rainbow (or Kamloops) trout tend to overshadow all other game species sought by anglers. They make up more than 60 per cent of the catch on a province-wide basis. Rainbow trout are an even larger part of the freshwater angler's take in the Nicola Valley.

Serious fishermen have found that the best angling is in the lakes to the north of Highway 8 and east of Highway 5. These lakes are managed largely by stocking and habitat improvement. In a few selected lakes, fishing restrictions are being used as a management tool. The British Columbia Sport Fishing regulations should be consulted before angling in Coquihalla Country.

•••

48

History

Early Native History

Interior Salish Indians have inhabited British Columbia for approximately 7,000 to 9,000 years. They were initially year-round nomads, and it was not until about 3,000 years ago that they began spending the winters in pit-house communities.

Evidence of pit-houses remain in the form of circular depressions at their former campsites near major lakes and streams. The pit-house, or kekuli, was constructed by excavating a circular pit one metre deep and up to 10 metres in diameter. Sturdy posts were set near the centre to support a framework of logs and poles which was then covered with soil. An opening at the top of the dome-shaped home served as a chimney and entrance. Kekuli depressions can still be seen at Monck Park, on the north shore of Nicola Lake.

James A. Teit, a leading Indian authority and resident of Merritt at the time of his death in 1922, suggested in his writing that the earliest permanent inhabitants of the Nicola Valley were a small tribe of Athabaskans. These Natives were closely related to the Chilcotins, and may have migrated southward from the Mackenzie River region. They occupied almost all of the Nicola and Similkameen valleys but were soon challenged by the Thompson and Okanagan Salish Indians.

According to Teit, the Salish had originally migrated northward from what is now the northwestern U.S.A. The Thompson Salish main settlement was at Lytton. As their numbers grew, they moved up the Thompson River and Nicola River, establishing communities throughout most of the lower Nicola and upper Similkameen valleys. The Okanagan Salish appear to have spread from the south Okanagan into the lower Similkameen and, via the north Okanagan, into the upper reaches of the Nicola River on the Douglas Plateau.

The Athabaskans fought the Okanagans, but were eventually absorbed by the Thompson Salish, losing their language and identity.

At the time of the arrival of the first fur traders, the main Indian communities were at the present site of Merritt; at the foot of Nicola Lake; near Quilchena; and at Douglas Lake.

Alexander Ross, then under the employ of the U.S. based Pacific Fur Company, is believed to have been the first white trader to visit the Nicola Valley. Ross and a companion were returning to Fort Okanogan in January 1813, following a Christmas visit to their previously established post at what is now Kamloops. They ran into difficulty with deep, crusted snow on the plateau country. After several days of very slow travel, they were assisted by the Natives who wintered near Douglas and Nicola lakes.

For the next half century, fur traders from the Pacific Fur Company, the North West Company, and the Hudson's Bay Company successively maintained an amicable, though sometimes paternal, relationship with Natives of the district.

Among the most powerful of the Nicola Valley Natives that the fur traders dealt with was Walking Grizzly Bear (Hwistesmetxeqen) whom they referred to by the more manageable name of Nicholas. The natives had difficulty with Nicholas and shortened it to N'kuala or Nicola. Chief Nicola, an Okanagan Salish from Douglas Lake, gained the respect of both the Natives and the whites. His name appears repeatedly on maps of the period, suggesting that he had influence over most of the region from the Thompson River to the Okanagan Valley. Nicola's name lives in the valley's largest lake and main river.

The Brigade Trails

The Hudson's Bay Company horse brigades must have been a magnificent sight as they filed through the Nicola Valley and over the mountain passes of the North Cascades. History tells us that from the first light of day until the heat of the afternoon sun, the trail was alive with the sounds of snorting horses and plodding hooves. The smell of hard-earned sweat permeated the air and the heavy diamond-hitched loads swayed in cadence with the songs of the voyageurs. A traveller of the day would certainly have stood in awe at the sight of 400 horses and 50 men winding along the trail.

Leading the procession was the guide, the trader or factor, and the clerks. At a respectful distance followed a white lead mare, adorned with a bell. Packers, frequently Natives, spread themselves out along the line of horses, each in charge of five to ten animals. At the rear was a rag-tag band that often included packers' families and other travellers.

For half a century, the brigade trails were the routes of trade and commerce in the interior of British Columbia. The waterways that served the fur traders in eastern Canada were not navigable in this land of high mountains and wild rivers. Horses and men were the sole movers of furs and trade goods in southwestern British Columbia from 1811 until the 1860s.

The colorful history of the fur brigades began in 1811 when Astorians David Stuart and Ovid de Montigny, of the New York based Pacific Fur Company, entered what is now British Columbia. They travelled north from Fort Okanogan, on the Columbia River, to the junction of the North and South Thompson rivers. Where the city of Kamloops stands today, they spent the winter among a large community of Shuswap Indians. In the following year, they were joined by Alexander Ross, and established Shuswap Fort, an American fur trading post well within Canada's present boundaries. In the same year, the Montreal-based North West Company established Thompson's River Post nearby. The two rival companies coexisted amicably until the Nor'Westers bought out the Astorians in 1813.

It was not until after the merger of the North West Company and the Hudson's Bay Company in 1821 that heavy use of the brigade routes was made. The route north of Kamloops followed the east side of the North Thompson River to Little Fort before striking westward across the upland lake country. From Horse Lake, near 100 Mile House, the trail wound northward following the approximate route of Highway 97 to Williams Lake, then north to the Fraser River at Fort Alexandria. This was the south end of the upper Fraser River navigation, and from Fort Alexandria the canoe became the mode of transportation to Fort George and Fort St. James. South of Kamloops, the horse brigades followed the South Thompson River up to Monte Creek and then proceeded south to Okanagan Lake along the approximate route of Highway 97. They then followed the western slopes of the Okanagan Valley until they reached the Columbia River.

In 1842, Fort Kamloops was erected on the west side of the North Thompson River and a new route established to the north country. It followed the north side of Kamloops Lake to Deadman Creek, then proceeded northward via Brigade Creek, Loon Lake and Green Lake, joining the earlier trail at the west end of Horse Lake. This continued to be an important transportation artery until the Cariboo Wagon Road was completed in the mid-1860s.

The Oregon Treaty, signed in 1846, established the 49th Parallel as the boundary between Canada and the United States. While the treaty settled a long-standing international dispute, it created a whole new set of problems for the Hudson's Bay Company. Although the treaty allowed free navigation of the Columbia River, the Hudson's Bay Company, long a law unto themselves, found Fort Vancouver, their main establishment, under foreign control. Indian difficulties, border disputes and an irksome duty soon persuaded The Company that a new route to the Pacific was absolutely essential.

Alexander Caulfield Anderson was assigned the task of finding a route from Fort Kamloops to the navigable waters of the lower Fraser River. After rejecting a western bypass of the Fraser Canyon via Seton, Anderson and Harrison lakes in 1846, Anderson sought a route that roughly paralleled the present Hope-Princeton Highway (Highway 3) for 40 kilometres and then headed north across the Cascade Mountains at Snass Creek.

However, when Chief Factor James Douglas did not feel this route was feasible, Anderson began mapping a route from the head of lower Fraser River navigation at Yale, up the west side of the river to Spuzzum. Here he crossed the river, then began a steep climb up the mountainside a short distance north of Alexandra. His trail took him over the cliffs of the Fraser Canyon and down to the Anderson River which he followed upstream to Uztlius Creek before crossing the divide into the Nicola Watershed. After crossing Maka Creek, Anderson's trail dropped into the Coldwater Valley just south of Midday Creek. He followed the Coldwater River downstream to its junction with the Nicola River, where Merritt stands today. From this point he followed the general route of Highway 5A to Fort Kamloops.

More Indian difficulties south of the border forced the Hudson's Bay Company to abandon the Columbia River posts. Ovid Allard and his men established Fort Yale in the spring of 1848, in anticipation of the first brigades to use the new Anderson trail. The arrival of the outgoing brigade of 1848 was not a time of the usual joy and celebration. The trip from the Forks (Merritt) had taken 10 days, during seven of which the horses were without food. The crossing of the Fraser near Spuzzum was a near disaster with many of the horses swept away by the boiling current. The incoming brigade of 1848 was even more difficult. The men fought a hard battle to get the canoes up the Fraser to Fort Yale. The river crossing at Spuzzum was even worse than on the previous trip partly because of the freshet and partly because hundreds of curious Natives interrupted their salmon fishing to "inspect" this strange group.

Douglas ordered the route abandoned, and work started on a trail explored by Henry Peers in 1848. In the spring of 1849, a fur trading post was established at the mouth of the Coquihalla River, now the town of Hope. While construction was underway, the outward brigade made the last trip over the Anderson trail.

Leaving the Fraser at Fort Hope, the inward Brigade of 1849 headed almost due east, avoiding the lower canyon of the Coquihalla and then following the river upstream until the junction with Peers Creek. The trail stayed on the north side of Peers Creek, following it up to the headwaters before making the steep climb to the top of Manson Ridge. The descent through Fools Pass and down to the Sowaqua was murderous and claimed many an errant horse. After crossing the Sowaqua, the trail climbed up and over 1,850-metre-high Mount Davis. The steep, rocky ascent from Campement du Chevreuil to Palmer Pond on Mount Davis was probably the worst of the trip. From Palmer Pond, the trail descended into the headwaters of Podunk Creek, following it to the Tulameen River. The brigade avoided the worst of the Tulameen Canyon by following Blackeyes' Trail across Lodestone Mountain to Campement des Femmes, near the village of Tulameen.

The horse brigade had conquered the Cascade Mountains. Another day's travel north through the Otter Valley and they were into the Nicola Watershed near Aspen Grove. The route along the west side of Quilchena Creek was easy compared to the Cascades, and when they reached Nicola Lake they were well into the open grasslands. They climbed out of the valley floor near Napier Lake, angling to the west and passing through the open country near Brigade and McLeod lakes before descending into the Thompson Valley at Kamloops.

From the first outward brigade of 1848 until the completion of the Harrison-Lillooet Road and the Cariboo Wagon Road, the brigade trails through the Nicola Valley were an important route of commerce. Fur traders, gold seekers, politicians and farmers used these corridors through the mountains and plateau country. Such famous men as A.C. Anderson, Sir James Douglas and Sir Matthew Baillie Begbie passed this way on their journey into the history of British Columbia.

Although numerous mentions of the trails have appeared in books, magazines and newspapers since the last brigades plodded over the mountains, it is only since the 1960s that serious effort has been made to relocate the old trails. Much of this effort has been expended in the North Cascade Mountains between Hope and Tulameen where the some of the axe blazes of more than a century ago remain.

The trail from the headwaters of Peers Creek to the top of Lodestone Mountain is now marked. The trip, a challenge to even the most experienced backpacker, takes as much as a week and is best travelled from mid-July to mid-September with an experienced guide.

The Okanagan Similkameen Parks Society (P.O. Box 787, Summerland, BC VOH IZO) is one of several organizations that has worked hard to have the old trails of the North Cascade Mountains preserved as part of our national heritage.

Settlement

Details are sketchy, but it is generally believed that Jesus Garcia, a Mexican packer on his way to the goldfields of the Cariboo, was the first to bring cattle and sheep into the Nicola Valley. Garcia brought his herd through the perilous Coquihalla Pass in 1860, and spent the winter at Spanish Springs, now Godey Creek, on the east boundary of the city of Merritt.

When the first permanent settlers drifted into the valley in the late 1860s, they quickly realized the agricultural potential. Hardy pioneers such as John Clapperton and Edwin Dalley helped establish a thriving community at the foot of Nicola Lake. Unlike many of the gold rush towns that these men had passed through, Nicola became a community of substance. Stores, livery stables, a school and a church were constructed. As the community grew, a sawmill and flour mill became the basis for local industry.

In 1882, Michael Hagan, then editor of the Yale-based Inland Sentinel, reported visiting Nicola on a return trip from Kamloops. "...we were at Fenson's Mills, and here it was we beheld the most business-looking place after leaving Kamloops. The mill power for both the grist mill and saw mill, where also a planer and shingle arrangement were to be seen, was from a fine stream running from the Nicola River."

Elsewhere in the valley, small settlements were springing up at Lower Nicola, Quilchena, Moore's (at the north end of Nicola Lake) and Forksdale (now Merritt).

Transportation was a problem in the early days. In 1874, the local residents lobbied the British Columbia Lieutenant Governor for a road: "The undersigned settlers of the Kamloops, Okanagan, Nicola and Cache Creek Valleys, beg to petition Your Honor, for the construction of a road from the South end of Nicola Forks, up the Coldwater Valley to the summit of the Coquhalla [sic], thence down the Coquhalla to Fort Hope. The distressed condition of the stockraisers of the district, owing to their having no outlet by which they can drive to the now almost only beef market in the Province, together with the fact that the cattle ranges are becoming overstocked and destroyed, we trust will induce you to make some efforts for our relief."

The decade of the seventies was a period of road building in the Nicola Valley. The wagon road from Spences Bridge was completed in 1875. In 1876, the ranchers' pleas were heeded and the Hope Trail was built through the Coldwater Valley and Coquihalla Canyon, improving local access to the growing beef markets of the lower mainland. The wagon road from Kamloops to Nicola was built in 1877-78, opening the ranch country to the north to carriage traffic.

The construction of the Canadian Pacific Railway in the 1880s provided a vast new market for Nicola Valley beef. This sparked the growth of the Douglas Lake Cattle Company with, as its nucleus, land originally homesteaded by John Douglas in 1872. Today the Douglas Lake Ranch claims to be the largest in Canada, running 15,000 Hereford cattle on 222,000 hectares of grasslands, lakes and timber.

Dan McGinnis, a young man of 19 when he trekked into the valley in June, 1883, wrote of the hardships faced by early ranchers in getting their cattle to the railway crews and lower mainland markets.

"I made two trips a month, taking 150 head each trip," wrote McGinnis. "Accompanied by five Indians, I would drive the cattle through the Coquihalla Pass to Hope. This trail was barely wide enough for the cattle to walk in single file. On one side rose a wall of rock, and on the other side was a sheer drop of 300 feet. On this trail was a narrow bridge of three logs, off which many of the cattle fell, and I lost as many as eleven steers in one trip."

The first mention of Nicola Valley coal appeared in the New Westminster British Columbian in 1869, where it was stated that blacksmiths on the Cariboo Wagon Road were using valley coal brought out by packhorse. Test holes were drilled near the confluence of the Nicola and Coldwater rivers in 1892-93. The results were "far beyond expectations" and lobbying for a railway into the valley increased in tempo.

In the early 1900s, William Hamilton Merritt held a government charter for the Nicola, Kamloops and Similkameen Coal and Railway Company. W.H. Merritt had the authority to construct a rail line from Chilliwack to Spences Bridge via the Coquihalla, Coldwater and Nicola valleys. A townsite named Forksdale was surveyed before W.H. Merritt's plans fell through.

In 1904, John Hendry staked 1,100 hectares of coal rights southwest of Forksdale with the intentions of starting a colliery. Hendry had already made his fortune in Vancouver as owner of Hastings Mills, and it was said he had good connections with the Canadian Pacific Railway. When the CPR extension was completed into the valley in 1907, the coal mining community of Forksdale was renamed in honor of W.H. Merritt.

A report in a July 1909 issue of Saturday Sunset summed up the situation. "Merritt, B.C., a new town in the Nicola Valley has a population of 500-800 souls. It has seven stores, three hotels, a railway, a couple of coal mines, livery stables, billiard parlor, cafes, opera house, bank and newspaper, but it has neither a church nor a railway station."

A church was still in the future, but work was started on the railway station the following month.

This was the era of coal and the steam locomotive, and railroad construction continued across southern British Columbia. The search for a route for a Coast-to-Kootenay railway led to renewed interest in the Coquihalla region. After numerous surveys and false starts, the Kettle Valley Railway started carving a railway right-of-way down the Coquihalla Canyon. Andrew McCulloch is justly credited with the greatest feat of railway engineering up to that time, building 32 bridges in 65 kilometres.

"Six of the bridges are steel, three of them are bridges that are as big as any in this country," said a 1915 report in the Merritt Herald. "The remaining 26 bridges are to be built of wood and over six million feet of timber will be used."

In 1916, after almost two decades of construction delays, the last spike was driven on the famous Kettle Valley Railway. For more than four decades, the Coast-to-Kootenay rail line faced the snow and ice of the Coquihalla Canyon. The Coquihalla link was finally broken by a giant mud slide in 1959, the last shot in the battle against Nature and declining revenues. A portion of the renowned KVR remained a CPR spur line linking Penticton with the main line at Spences Bridge until May, 1989 when the last lumber train rolled down the tracks. (At the time of writing, a Rails to Trails group was lobbying to have the railway right-of-way converted to a recreational corridor.)

The first half of the twentieth century saw many changes in the Nicola Valley. The coal mines drew the centre of commerce away from Nicola, and Merritt grew at its expense. Then coal gradually lost favour as an energy source and highways replaced railways as the main transportation routes. These changes provided the occasional severe setback, but the valley did not die. Instead, lumbering, copper mining, ranching and tourism have all contributed to a steady, orderly growth.

Today Merritt is still the commercial centre of the Nicola Valley. Some of the 19th century settlements have disappeared while others, such as Nicola, have changed little. Landmarks including the Murray Church at Nicola, the Coldwater Hotel at Merritt, and the Quilchena Hotel serve as reminders of a colorful past: a past that is relived each Labour Day weekend in the rip-roaring celebration of the Nicola Valley Rodeo.

•••

49

Bibliography

Books

AKRIGG, G.P.V. & Helen B. *1001 British Columbia Place Names* (Discovery Press, Vancouver, BC 1973).

AKRIGG, G.P.V. & Helen B. *British Columbia Chronicle, 1847-1871* (Discovery Press, Vancouver, BC 1977).

BULMAN, T. Alex. *Kamloops Cattlemen*, (Gray's Publishing, Sidney, BC 1972).

HARRIS, Bob, Harley Hatfield and Peter Tassie. *The Okanagan Brigade Trail* (Vernon, BC 1989 ISBN 0-9694207-0-6).

MILLIKEN, A.C. "Paul Fraser's Grave". *BC Digest* (April, 1962) pp 59-60.

NICOLA VALLEY MUSEUM ARCHIVES ASSOCIATION *Merritt & the Nicola Valley: An Illustrated History.* (Sonotek Publishing Ltd., Merritt, BC 1989/1998 ISBN 0-929069-01-3).

READ, Stanley E. *A Place Called Pennask*, (Mitchell Press, Vancouver, BC 1977).

ROTHENBURGER, Mel. *'We've Killed Johnny Ussher!'* (Mitchell Press, Vancouver, BC 1973).

SANFORD, Barrie *McCulloch's Wonder — The Story of the Kettle Valley Railway*. (Whitecap Books, West Vancouver, BC 1977).

SHEWCHUK, Murphy O. *Backroads Explorer: Similkameen and South Okanagan.* (Hancock House, Surrey, BC 1988).

SHEWCHUK, Murphy O. *Backroads Explorer: Thompson-Cariboo* (Maclean Hunter, Vancouver, BC 1985).

SHEWCHUK, Murphy O. *The Craigmont Story* (Hancock House, Surrey, BC 1983).

SHEWCHUK, Murphy O. *Exploring the Nicola Valley* (Douglas & McIntyre, Vancouver, BC 1981).

SHEWCHUK, Murphy O. *Fur, Gold and Opals* (Hancock House, Surrey, BC 1975).

SHEWCHUK, Murphy O. *Okanagan Country: An Outdoor Recreation Guide* (Sonotek Publishing Ltd., 1992 ISBN 0-929069-04-8).

SMITH, Jessie Ann *Widow Smith of Spence's Bridge* (Sonotek Publishing Ltd., Merritt, BC 1989 ISBN 0-929069-00-5).

WOOLLIAMS, Nina G. *Cattle Ranch - The Story of the Douglas Lake Cattle Company* (Douglas & McIntrye, Vancouver BC 1979).

•••

Additional information

Updated information and corrections to this book will be posted on the World Wide Web at <http://www.sonotek.com>. Please e-mail any suggestions or comments to <publisher@sonotek.com>.

Merritt Forest District Recreation Sites Map, available from Ministry of Forests, Bag 4400, Merritt, BC V1K 1B8 or from their office at Highway 5A and Airport Road, Merritt, BC Telephone (250) 378-8400.

For topographical maps, contact the Nicola Valley Museum Archives, 2202 Jackson Avenue, PO Box 1262, Merritt, BC V1K 1B8. Telephone (250) 378-4145.

Provincial Parks information is available from the Ministry of Lands, Parks and Housing, Parks and Outdoor Recreation Division, 1019 Wharf Street, Victoria, BC V8W 2Y9, or from BC Parks, Fraser Valley District, Box 10, Cultus Lake, BC V0X 1H0 or BC Parks, Room 101, 1050 West Columbia St., Kamloops, BC V2C 1L4.

Hope Travel Infocentre, 919 Water Avenue, PO Box 370, Hope, BC V0X 1L0. Telephone (604) 869-2021, Fax (604) 869-2160.

Kamloops Travel Infocentre, 10 - 10th Avenue, Kamloops, BC V2C 6J7. Telephone (250) 374-3377.

Kelowna Travel Infocentre, 544 Harvey Avenue, Kelowna, BC V1Y 6C9. Telephone (250) 861-1515.

District of Logan Lake Travel Infocentre, Box 190, Logan Lake, BC V0K 1W0. Telephone (250) 523-6225.

Merritt & District Chamber of Commerce, P.O. Box 1649, Merritt, BC V1K 1B8. Telephone (250) 378-5634, Fax (250) 378-6561.

•••

50

Index

C

D

N

O

P

About the Author

MURPHY SHEWCHUK has been writing and illustrating articles and books since the mid-1960s.

Born in Hamilton, Ontario, in 1943, he grew up in the British Columbia mining town of Pioneer Gold Mines and developed a triple interest in outdoor exploring, photography and electronics while still a teenager.

After a stint in the RCAF in eastern Canada, he moved back to BC to a job in electronics with writing and photography as his major interests. In 1971, his work took him to Kamloops, BC, where he began a weekly *Outdoor Scene* column for the *Kamloops Sentinel.* From the column, he moved to features in *BC Outdoors* and many other magazines. He now lives with his wife Katharine in Merritt, BC.

Figure 51
Murphy O. Shewchuk

More than 300 of his magazine articles and 1000 of his photographs have appeared in such publications as *Adventure Travel, BC Outdoors, Camping Canada, Canadian Geographic, Field & Stream, MotorHome, Photo Life, Skyword, Western Living, and Westworld.*

He has also written eight other books including: *Exploring Kamloops Country; Fur, Gold & Opals; Exploring the Nicola Valley; Cariboo; The Craigmont Story; Backroads Explorer Vol 1: Thompson-Cariboo; Backroads Explorer Vol 2: Similkameen & South Okanagan* and *Okanagan Country: An Outdoor Recreation Guide.*

In addition to his life-long interest in photography and exploring the mountains of western Canada, Murphy is a workshop speaker at writer's conferences across Canada and in the U.S.A. His writing and photography have received awards from the Outdoor Writers of Canada and the Macmillan Bloedel newspaper journalism competitions. In 1996, Shewchuk was elected President of the Canadian Authors Association.

•••